Cadillac Allanté
1986-1993

Compiled by
R.M.Clarke

ISBN 1 85520 3324

BROOKLANDS BOOKS LTD.
P.O. BOX 146, COBHAM,
SURREY, KT11 1LG. UK

BROOKLANDS BOOKS

BROOKLANDS ROAD TEST SERIES

Abarth Gold Portfolio 1950-1971
AC Ace & Aceca 1953-1983
Alfa Romeo Giulietta Gold Portfolio 1954-1965
Alfa Romeo Giulia Berlinas 1962-1976
Alfa Romeo Giulia Coupés 1963-1976
Alfa Romeo Giulia Coupés Gold P. 1963-1976
Alfa Romeo Spider 1966-1990
Alfa Romeo Spider Gold Portfolio 1966-1991
Alfa Romeo Alfasud 1972-1984
Alfa Romeo Alfetta Gold Portfolio 1972-1987
Alfa Romeo Alfetta GTV6 1980-1986
Allard Gold Portfolio 1937-1959
Alvis Gold Portfolio 1919-1967
AMX & Javelin Muscle Portfolio 1968-1974
Armstrong Siddeley Gold Portfolio 1945-1960
Aston Martin Gold Portfolio 1972-1985
Aston Martin Gold Portfolio 1985-1995
Audi Quattro Gold Portfolio 1980-1991
Austin A30 & A35 1951-1962
Austin Healey 100 & 100/6 Gold P. 1952-1959
Austin Healey 3000 Gold Portfolio 1959-1967
Austin Healey Sprite 1958-1971
Barracuda Muscle Portfolio 1964-1974
BMW Six Cylinder Coupés 1969-1975
BMW 1600 Collection No.1 1966-1981
BMW 2002 Gold Portfolio 1968-1976
BMW 316, 318, 320 (4 cyl.) Gold P. 1975-1990
BMW 320, 323, 325 (6 cyl.) Gold P. 1977-1990
BMW M Series Performance Portfolio 1976-1993
BMW 5 Series Gold Portfolio 1981-1987
Bricklin Gold Portfolio 1974-1975
Bristol Cars Gold Portfolio 1946-1992
Buick Automobiles 1947-1960
Buick Muscle Cars 1965-1970
Cadillac Allanté 1986-1993
Cadillac Automobiles 1949-1959
Cadillac Automobiles 1960-1969
Charger Muscle Portfolio 1966-1974
Chevrolet 1955-1957
Chevrolet Impala & SS 1958-1971
Chevrolet Corvair 1959-1969
Chevy II & Nova SS Muscle Portfolio 1962-1974
Chevy El Camino & SS 1959-1987
Chevelle & SS Muscle Portfolio 1964-1972
Chevrolet Muscle Cars 1966-1971
Chevy Blazer 1969-1981
Chevrolet Corvette Gold Portfolio 1953-1962
Chevrolet Corvette Sting Ray Gold P. 1963-1967
Chevrolet Corvette Gold Portfolio 1968-1977
High Performance Corvettes 1983-1989
Camaro Muscle Portfolio 1967-1973
Chevrolet Camaro Z28 & SS 1966-1973
Chevrolet Camaro & Z28 1973-1981
High Performance Camaros 1982-1988
Chrysler 300 Gold Portfolio 1955-1970
Chrysler Valiant 1960-1962
Citroen Traction Avant Gold Portfolio 1934-1957
Citroen 2CV Gold Portfolio 1948-1989
Citroen DS & ID 1955-1975
Citroen DS & ID Gold Portfolio 1955-1975
Citroen SM 1970-1975
Cobras & Replicas 1962-1983
Shelby Cobra Gold Portfolio 1962-1969
Cobras & Cobra Replicas Gold P. 1962-1989
Cunningham Automobiles 1951-1955
Daimler SP250 Sports & V-8 250 Saloon Gold P. 1959-1969
Datsun Roadsters 1962-1971
Datsun 240Z 1970-1973
Datsun 280Z & ZX 1975-1983
DeLorean Gold Portfolio 1977-1995
Dodge Muscle Cars 1967-1970
Dodge Viper on the Road
ERA Gold Portfolio 1934-1994
Excalibur Collection No.1 1952-1981
Facel Vega 1954-1964
Ferrari Dino 1965-1974
Ferrari Dino 308 1974-1979
Ferrari 328 • 348 • Mondial Gold Portfolio 1986-1994
Fiat 500 Gold Portfolio 1936-1972
Fiat 600 & 850 Gold Portfolio 1955-1972
Fiat Pininfarina 124 & 2000 Spider 1968-1985
Fiat-Bertone X1/9 1973-1988
Fiat Abarth Performance Portfolio 1972-1987
Ford Consul, Zephyr, Zodiac Mk.I & II 1950-1962
Ford Zephyr, Zodiac, Executive, Mk.III & Mk.IV 1962-1971
Ford Cortina 1600E & GT 1967-1970
High Performance Capris Gold Portfolio 1969-1987
Capri Muscle Portfolio 1974-1987
High Performance Fiestas 1979-1991
High Performance Escorts Mk.I 1968-1974
High Performance Escorts Mk.II 1975-1980
High Performance Escorts 1980-1985
High Performance Escorts 1985-1990
High Performance Sierras & Merkurs
Gold Portfolio 1983-1990
Ford Automobiles 1949-1959
Ford Fairlane 1955-1970
Ford Ranchero 1957-1959
Ford Thunderbird 1955-1957
Ford Thunderbird 1958-1963
Ford Thunderbird 1964-1976
Ford GT40 Gold Portfolio 1964-1987
Ford Bronco 1966-1977
Ford Bronco 1978-1988
Holden 1948-1962
Honda CRX 1983-1987
International Scout Gold Portfolio 1961-1980
Isetta 1953-1964
Iso & Bizzarrini Gold Portfolio 1962-1974
Jaguar and SS Gold Portfolio 1931-1951
Jaguar XK120, 140, 150 Gold P. 1948-1960
Jaguar Mk.VII, VIII, IX, X, 420 Gold P. 1950-1970
Jaguar Mk.1 & Mk.2 Gold Portfolio 1959-1969
Jaguar E-Type Gold Portfolio 1961-1971
Jaguar E-Type V-12 1971-1975
Jaguar XJ12, XJ5.3, V12 Gold P. 1972-1990
Jaguar XJ6 Series I & II Gold P. 1968-1979
Jaguar XJ6 Series III 1979-1986
Jaguar XJ6 Gold Portfolio 1986-1994
Jaguar XJS Gold Portfolio 1975-1988
Jaguar XJS Gold Portfolio 1988-1995
Jeep CJ5 & CJ6 1960-1976
Jeep CJ5 & CJ7 1976-1986
Jensen Cars 1946-1967
Jensen Cars 1967-1979
Jensen Interceptor Gold Portfolio 1966-1986
Jensen Healey 1972-1976
Lagonda Gold Portfolio 1919-1964
Lamborghini Countach & Urraco 1974-1980
Lamborghini Countach & Jalpa 1980-1985
Lancia Fulvia Gold Portfolio 1963-1976
Lancia Beta Gold Portfolio 1972-1984
Lancia Delta Gold Portfolio 1979-1994
Lancia Stratos 1972-1985
Land Rover Series I 1948-1958
Land Rover Series II & IIa 1958-1971
Land Rover Series III 1971-1985
Land Rover 90 110 Defender Gold Portfolio 1983-1994
Land Rover Discovery 1989-1994
Lincoln Gold Portfolio 1949-1960
Lincoln Continental 1961-1969
Lincoln Continental 1969-1976
Lotus Sports Racers Gold Portfolio 1953-1965
Lotus Seven Gold Portfolio 1957-1974
Lotus Caterham Seven Gold Portfolio 1974-1995
Lotus Elite 1957-1964
Lotus Elite & Eclat 1974-1982
Lotus Elan Gold Portfolio 1962-1974
Lotus Elan Collection No. 2 1963-1972
Lotus Elan & SE 1989-1992
Lotus Cortina Gold Portfolio 1963-1970
Lotus Europa Gold Portfolio 1966-1975
Lotus Elite & Eclat 1974-1982
Lotus Turbo Esprit 1980-1986
Marcos Cars 1960-1988
Maserati 1965-1970
Maserati 1970-1975
Mercedes 190 & 300 SL 1954-1963
Mercedes 230/250/280SL 1963-1971
Mercedes G Wagen 1981-1994
Mercedes Benz SLs & SLCs Gold P. 1971-1989
Mercedes S & 600 1965-1972
Mercedes S Class 1972-1979
Mercedes SLs Performance Portfolio 1989-1994
Mercury Muscle Cars 1966-1971
Messerschmitt Gold Portfolio 1954-1964
MG Gold Portfolio 1929-1939
MG TA & TC Gold Portfolio 1936-1949
MG TD &TF Gold Portfolio 1949-1955
MGA & Twin Cam Gold Portfolio 1955-1962
MG Midget Gold Portfolio 1961-1979
MGB Roadsters 1962-1980
MGB MGC & V8 Gold Portfolio 1962-1980
MGB GT 1965-1980
Mini Gold Portfolio 1959-1969
Mini Gold Portfolio 1969-1980
High Performance Minis Gold Portfolio 1960-1973
Mini Cooper Gold Portfolio 1961-1971
Mini Moke Gold Portfolio 1964-1994
Mopar Muscle Cars 1964-1967
Morgan Three-Wheeler Gold Portfolio 1910-1952
Morgan Plus 4 & Four 4 Gold P. 1936-1967
Morgan Cars 1960-1970
Morgan Cars Gold Portfolio 1968-1989
Morris Minor Collection No. 1 1948-1980
Shelby Mustang Muscle Portfolio 1965-1970
High Performance Mustang IIs 1974-1978
High Performance Mustangs 1982-1988
Nash-Austin Metropolitan Gold P. 1954-1962
Oldsmobile Automobiles 1955-1963
Oldsmobile Muscle Cars 1964-1971
Oldsmobile Toronado 1966-1978
Opel GT Gold Portfolio 1968-1973
Packard Gold Portfolio 1946-1958
Pantera Gold Portfolio 1970-1989
Panther Gold Portfolio 1972-1990
Plymouth Muscle Cars 1966-1971
Pontiac Tempest & GTO 1961-1965
Pontiac Muscle Cars 1966-1972
Pontiac Firebird & Trans-Am 1973-1981
High Performance Firebirds 1982-1988
Pontiac Fiero 1984-1988
Porsche 356 Gold Portfolio 1953-1965
Porsche 911 1965-1969
Porsche 911 1970-1972
Porsche 911 1973-1977
Porsche 911 Carrera 1973-1977
Porsche 911 Turbo 1975-1984
Porsche 911 SC & Turbo Gold Portfolio 1978-1983
Porsche 911 Carrera & Turbo Gold P. 1984-1989
Porsche 914 Gold Portfolio 1969-1976
Porsche 924 Gold Portfolio 1975-1988
Porsche 928 Performance Portfolio 1977-1994
Porsche 944 Gold Portfolio 1981-1991
Range Rover Gold Portfolio 1970-1985
Range Rover Gold Portfolio 1986-1995
Reliant Scimitar 1964-1986
Riley Gold Portfolio 1924-1939
Riley 1.5 & 2.5 Litre Gold Portfolio 1945-1955
Rolls Royce Silver Cloud & Bentley 'S' Series
Gold Portfolio 1955-1965
Rolls Royce Silver Shadow Gold P. 1965-1980
Rolls Royce & Bentley Gold P. 1980-1989
Rover P4 1949-1959
Rover P4 1955-1964
Rover 3 & 3.5 Litre Gold Portfolio 1958-1973
Rover 2000 & 2200 1963-1977
Rover 3500 1968-1977
Rover 3500 & Vitesse 1976-1986
Saab Sonett Collection No.1 1966-1974
Saab Turbo 1976-1983
Studebaker Gold Portfolio 1947-1966
Studebaker Hawks & Larks 1956-1963
Avanti 1962-1990
Sunbeam Tiger & Alpine Gold P. 1959-1967
Toyota MR2 1984-1988
Toyota Land Cruiser 1956-1984
Triumph TR2 & TR3 Gold Portfolio 1952-1961
Triumph TR4, TR5, TR250 1961-1968
Triumph TR6 Gold Portfolio 1969-1976
Triumph TR7 & TR8 Gold Portfolio 1975-1982
Triumph Herald 1959-1971
Triumph Vitesse 1962-1971
Triumph Spitfire Gold Portfolio 1962-1980
Triumph 2000, 2.5, 2500 1963-1977
Triumph GT6 Gold Portfolio 1966-1974
Triumph Stag 1970-1980
TVR Gold Portfolio 1959-1986
TVR Performance Portfolio 1986-1994
VW Beetle Gold Portfolio 1935-1967
VW Beetle Gold Portfolio 1968-1991
VW Beetle Collection No.1 1970-1982
VW Karmann Ghia 1955-1982
VW Bus, Camper, Van 1954-1967
VW Bus, Camper, Van 1968-1979
VW Bus, Camper, Van 1979-1989
VW Scirocco 1974-1981
VW Golf GTI 1976-1986
Volvo PV444 & PV544 1945-1965
Volvo Amazon-120 Gold Portfolio 1956-1970
Volvo 1800 Gold Portfolio 1960-1973
Volvo 140 & 160 Series Gold Portfolio 1966-1975

Forty Years of Selling Volvo

BROOKLANDS ROAD & TRACK SERIES

Road & Track on Alfa Romeo 1949-1963
Road & Track on Alfa Romeo 1964-1970
Road & Track on Alfa Romeo 1971-1976
Road & Track on Alfa Romeo 1977-1989
Road & Track on Aston Martin 1962-1990
R & T on Auburn Cord and Duesenburg 1952-84
Road & Track on Audi & Auto Union 1952-1980
Road & Track on Audi & Auto Union 1980-1986
Road & Track on Austin Healey 1953-1970
Road & Track on BMW Cars 1966-1974
Road & Track on BMW Cars 1975-1978
Road & Track on BMW Cars 1979-1983
R & T on Cobra, Shelby & Ford GT40 1962-1992
Road & Track on Corvette 1953-1967
Road & Track on Corvette 1968-1982
Road & Track on Corvette 1982-1986
Road & Track on Corvette 1986-1990
Road & Track on Datsun Z 1970-1983
Road & Track on Ferrari 1975-1981
Road & Track on Ferrari 1981-1984
Road & Track on Ferrari 1984-1988
Road & Track on Fiat Sports Cars 1968-1987
Road & Track on Jaguar 1950-1960
Road & Track on Jaguar 1961-1968
Road & Track on Jaguar 1968-1974
Road & Track on Jaguar 1974-1982
Road & Track on Jaguar 1983-1989
Road & Track on Lamborghini 1964-1985
Road & Track on Lotus 1972-1981
Road & Track on Maserati 1952-1974
Road & Track on Maserati 1975-1983
R & T on Mazda RX7 & MX5 Miata 1986-1991
Road & Track on Mercedes 1952-1962
Road & Track on Mercedes 1963-1970
Road & Track on Mercedes 1971-1979
Road & Track on Mercedes 1980-1987
Road & Track on MG Sports Cars 1949-1961
Road & Track on MG Sports Cars 1962-1980
Road & Track on Mustang 1964-1977
R & T on Nissan 300-ZX & Turbo 1984-1989
Road & Track on Pontiac 1960-1983
Road & Track on Porsche 1951-1967
Road & Track on Porsche 1968-1971
Road & Track on Porsche 1972-1975
Road & Track on Porsche 1975-1978
Road & Track on Porsche 1979-1982
Road & Track on Porsche 1982-1985
Road & Track on Porsche 1985-1988
R & T on Rolls Royce & Bentley 1950-1965
R & T on Rolls Royce & Bentley 1966-1984
Road & Track on Saab 1972-1992
R & T on Toyota Sports & GT Cars 1966-1984
R & T on Triumph Sports Cars 1953-1967
R & T on Triumph Sports Cars 1967-1974
R & T on Triumph Sports Cars 1974-1982
Road & Track on Volkswagen 1951-1968
Road & Track on Volkswagen 1968-1978
Road & Track on Volkswagen 1978-1985
Road & Track on Volvo 1957-1974
Road & Track on Volvo 1977-1994
R&T - Henry Manney at Large & Abroad
R&T - Peter Egan's "Side Glances"

BROOKLANDS CAR AND DRIVER SERIES

Car and Driver on BMW 1955-1977
Car and Driver on BMW 1977-1985
C and D on Cobra, Shelby & Ford GT40 1963-84
Car and Driver on Corvette 1956-1967
Car and Driver on Corvette 1968-1977
Car and Driver on Corvette 1978-1982
Car and Driver on Corvette 1983-1988
C and D on Datsun Z 1600 & 2000 1966-1984
Car and Driver on Ferrari 1955-1962
Car and Driver on Ferrari 1963-1975
Car and Driver on Ferrari 1976-1983
Car and Driver on Mopar 1956-1967
Car and Driver on Mopar 1968-1975
Car and Driver on Mustang 1964-1972
Car and Driver on Pontiac 1961-1975
Car and Driver on Porsche 1955-1962
Car and Driver on Porsche 1963-1970
Car and Driver on Porsche 1970-1976
Car and Driver on Porsche 1977-1981
Car and Driver on Porsche 1982-1986
Car and Driver on Saab 1956-1985
Car and Driver on Volvo 1955-1986

BROOKLANDS PRACTICAL CLASSICS SERIES

PC on Austin A40 Restoration
PC on Land Rover Restoration
PC on Metalworking in Restoration
PC on Midget/Sprite Restoration
PC on Mini Cooper Restoration
PC on MGB Restoration
PC on Morris Minor Restoration
PC on Sunbeam Rapier Restoration
PC on Triumph Herald/Vitesse
PC on Spitfire Restoration
PC on Beetle Restoration
PC on 1930s Car Restoration

BROOKLANDS HOT ROD 'MUSCLECAR & HI-PO ENGINES' SERIES

Chevy 265 & 283
Chevy 302 & 327
Chevy 348 & 409
Chevy 350 & 400
Chevy 396 & 427
Chevy 454 thru 512
Chrysler Hemi
Chrysler 273, 318, 340 & 360
Chrysler 361, 383, 400, 413, 426, 440
Ford 289, 302, Boss 302 & 351W
Ford 351C & Boss 351
Ford Big Block

BROOKLANDS RESTORATION SERIES

Auto Restoration Tips & Techniques
Basic Bodywork Tips & Techniques
Basic Painting Tips & Techniques
Camaro Restoration Tips & Techniques
Chevrolet High Performance Tips & Techniques
Chevy Engine Swapping Tips & Techniques
Chevy-GMC Pickup Repair
Chrysler Engine Swapping Tips & Techniques
Custom Painting Tips & Techniques
Engine Swapping Tips & Techniques
Ford Pickup Repair
How to Build a Street Rod
Land Rover Restoration Tips & Techniques
MG 'T' Series Restoration Guide
MGA Restoration Guide
Mustang Restoration Tips & Techniques
Performance Tuning - Chevrolets of the '60's
Performance Tuning - Pontiacs of the '60's

BROOKLANDS MILITARY VEHICLES SERIES

Allied Military Vehicles No.1 1942-1945
Allied Military Vehicles No.2 1941-1946
Complete WW2 Military Jeep Manual
Dodge Military Vehicles No.1 1940-1945
Hail To The Jeep
Land Rovers in Military Service
Military & Civilian Amphibians 1940-1990
Off Road Jeeps: Civ. & Mil. 1944-1971
US Military Vehicles 1941-1945
US Army Military Vehicles WW2-TM9-2800
VW Kubelwagen Military Portfolio 1940-1990
WW2 Jeep Military Portfolio 1941-1945
2985

CONTENTS

Page	Article	Publication	Date		
5	Cadillac Allanté	*Motor Trend*	June		1986
7	Return of the Cadillac Flagship	*Autosport*	Sept	25	1986
8	Atlantic Crossing	*Autocar*	Aug	27	1986
12	Cadillac Allanté Tees off Against Mercedes-Benz 560SL Comparison Test	*Road & Track*	Nov		1986
20	Italian Bloodline	*Sporting Cars International*	Dec		1986
22	Daydream Believer	*Modern Motor*	Jan		1988
24	Buick Reatta and Cadillac Allanté	*Road & Track*	Mar		1988
32	Cadillac Allanté Road Test	*Motor Trend*	Apr		1988
36	Five Exotic Convertibles Comparison Test	*Road & Track*	July		1988
44	'89 New Cars Cadillac Allanté	*Car and Driver*	Oct		1988
45	Cadillac Allanté	*Road & Track*	May		1989
46	Cadillac Allanté versus Mercedes 560SL Comparison Test	*Car and Driver*	Feb		1989
53	Cadillac Allanté Road Test	*Automobile Magazine*	Oct		1989
56	Cadillac Allanté Long Term Test	*Motor Trend*	Jan		1990
60	Cadillac Allanté	*Road & Track Special*	Mar		1990
64	Cadillac Allanté Road Test	*Automobile Magazine*	Oct		1990
69	Cadillac Allanté	*Road & Track Special*	Mar		1992
74	1993 Cadillac Allanté	*Automobile Magazine*	Feb		1992
77	Cadillac Allanté Long Term Test	*Motor Trend*	July		1991
78	1993 Cadillac Allanté Road Test	*Road & Track Special*	Aug		1992
83	Cadillac Connections	*Road & Track*	Sept		1992
84	Cadillac Allanté, Jaguar XJS and Mercedes 300SL Comparison Test	*Car and Driver*	July		1992
91	'Northstar' Allanté	*Car South Africa*	June		1992
94	Prime Rib and Sushi Comparison Test	*Modern Motor*	Jan		1993
100	Cadillac Allanté	*Road & Track Special*	Feb		1993

ACKNOWLEDGEMENTS

Our aim at Brooklands Books is to make available to motoring enthusiasts that printed material which would otherwise be hard for them to find. As time goes on, and the interesting new cars of yesterday become the sought-after classics of today, we are obliged to dig ever deeper into our archives to satisfy enthusiast interest.

We must of course acknowledge our debt to those who have granted us permission to reproduce their copyright material - and we are pleased to do so. Sincere thanks therefore go to the managements of *Autocar, Automobile Magazine, Autosport, Car and Driver, Car South Africa, Modern Motor, Motor Trend, Road & Track* and *Sports Car International*.

R M Clarke

The Cadillac Allanté really should have been a success, but somehow it never realised its makers' expectations. Cadillac's intention was to carve for themselves a slice of the exotic sports car market then dominated by the SL Mercedes and the Jaguar XJ-S, and to that end they called in the great Italian stylist Pininfarina to design the body and put together the best mechanical specification they could muster.

The new car was introduced in the late summer of 1986 as a 1987 model. It looked good, but somehow the reality never quite lived up to the promise. The 170bhp V8 was no ball of fire and the interior was just too plasticky to offer serious competition to the well-established sportsters from Europe. Sales in that first season were a disappointing 3,363 cars; and the next year they actually dropped, to just 2,569.

So Cadillac tried again, fitting an enlarged 200bhp engine for the 1989 season, improving the passenger compartment and reworking the suspension. That was certainly better. For 1990 came ABS and a year later the car got Cadillac's highly-acclaimed new North Star V8 engine with 290bhp.

But the damage had been done. The car had already gained itself a reputation as a bland luxury cruiser and no amount of improvements were going to make it a real competitor for the (by then) much-improved European sports cabriolets. So Cadillac decided to cut their losses, and the car was withdrawn from production in 1993. As the road tests in this book show, it was a good car by the end, but it had always been just that little bit too controversial . . . . .

James Taylor

FIRST LOOK

1987 Cadillac Allanté

Pininfarina designs a 2-seater for the ultra-lux wars

by Daniel Charles Ross
PHOTOGRAPHY BY ED SPERKO, G.M. PHOTOGRAPHIC

Picture the luxury car market as Cadillac does—a pyramid. Draw a line across the base at about the one-quarter mark and label it the "near luxury" segment. Scribe another line just below the pyramid's tip, and label the large middle segment "luxury." In the resulting apex of this world according to Cadillac, the pinnacle of automotive achievement, pencil in the word "ultra-luxury."

It's this tiny segment of the luxury car market Cadillac will enter for the first time

Cadillac's Allanté is better looking in the metal than in any photograph

ever with the automobile displayed here, the '87 Cadillac Allanté. With this car, Cadillac enters the rarified air at the tip of the market pyramid and simultaneously demonstrates a bold commitment to change as never before in its history. This has to be an exciting time to work at Cadillac's Clark Street headquarters.

The Allanté tips the scales at about 3490 lb, a solid upper-middleweight with good reason. "It has the structural integrity of a closed coupe," a highly placed Cadillac source told us. "We weren't going to do a convertible unless we got that. So we had to put structure back into the vehicle" to ensure torsional rigidity, a facet of rag-top life not as closely scrutinized in some other convertibles.

The car is even better looking in the metal than it is in these photographs. Its galvanized steel body panels stretch to an overall length of 178.6 in. and sit astride a wheelbase of 99.4 in. Height with the hard top installed is a relatively spare 52.2 in. Steering is via power-assisted rack and pinion. The suspension is independent fore and aft, and consists primarily of MacPherson struts with a 33mm stabilizer bar forward, Mac-Struts and Eldo/Seville's composite leaf spring mounted transversely in the rear. The geometry began life as that used for Eldorado/Seville, and was fine-tuned with special rates for the power of the new car.

As reported previously in these pages, Allanté will be powered by Cadillac's HT4100 4.1-liter V-8, mounted transversely and driving 15 x 7.0-in. forged aluminum front wheels shod with P225/60VR15 Goodyear Eagles. In the Italo-American Allanté, however, the engine gets an Allanté-specific induction system, lower-restriction exhaust system, and roller lifters. Instead of the two-injector throttle body used in other Cadillacs, Allanté gets tuned intake runners and eight injectors, electronically fired in a sequential pattern. Compression ratio is reported as 8.5:1, but might go even higher. According to our highly placed source, expect the Allanté engine to make about 170 hp at 4400 rpm, with max torque of 240 lb-ft coming on by 3200 rpm. This configuration in testing has reportedly been good for high nines between zero and 60 mph.

Being able to accelerate like no other Cadillac in decades necessitates a requirement to stop with similar alacrity. Cadillac is deploying a Bosch anti-skid Electronic Braking System (EBS) on the Allanté, another first. It will modulate brake pressure to power disc brakes sized 10.25 x 1.0 in. front and 10.50 x 0.5 in. rear.

Putting those thoroughbreds to the pavement is a 4-speed automatic transmission dubbed the F7, a beefier version of GM's THM440 reportedly better able to handle the decidedly un-Cadillac-like power output. Intriguing about this transmission is a feature Cadillac calls Torque Management. "It's a computer that does for the automatic what the brain and a clutch do for a manual."

The Cadillac Allanté isn't intended to compete with the Porsches, Bimmers, and Mercedes-Benzes already crowding each other in the ultra-luxury market segment, so we frankly don't expect the car to outperform those competitors. But we do expect the Allanté to accelerate well, handle and brake like no other Cadillac product in history, and be pretty expensive. [MT]

Cadillac's elegant new Allanté convertible looks equally handsome with its steel hardtop bolted in place.

CADILLAC ALLANTE

ROAD CAR
By MIKE McCARTHY

For a two-seater, it's a large automobile. Big and plush, and very American in its character, the Allante is to challenge the Mercedes SL.

Return of the Cadillac flagship

They won't admit it out loud, but you can't help the feeling that the good folks at Cadillac are a mite worried about image.

The Diesel Seville didn't go down all that well, and their current base model is the Cimarron, a Vauxhall Cavalier-size car that is tiny by Cadillac's standards. What was needed was a flagship, a true Cadillac in the traditions of the V16s of the thirties, or the Eldorado dropheads of the fifties – the sort of car which appealed to those who had it made in a big way.

The 'in' car over there at the moment for the well-heeled over-40s is the Mercedes-Benz SL drophead, so something that would offer a viable alternative, with a Cadillac badge, would fit the bill. Of course, the Merc had the cachet of being built in Germany, so the other thing that was needed was a designer label, in effect. Add Pininfarina to the badging and you should be there.

Enter the Cadillac Allante.

The Allante is aimed straight at the Merc SL but with good ol' fashioned Yankee input, which means that luxury and refinement over-ride performance and roadholding or handling. You can see their point, actually: the Merc has to satisfy European standards of road manners as well as appeal to American criteria in the looks and comfort stakes. The Allante simply has to appeal to rich, middle-aged Americans who have to live with the double-nickel speed limit and the local country club. The result is a car soft by European standards but with more gadgets, a better ride and sheer refinement that is at least the match of the Merc – and, come to that, close to that of the XJS Jaguar.

The idea of a 4.1-litre V8 driving the front wheels is enough to make even the hardiest of front-wheel drive enthusiasts blanch but, what with emission controls and other power-sapping devices, it only produces 170bhp at a low 4300rpm. Combined with an all-up weight of about 32cwt, this gives the sort of power to weight ratio that isn't excessive for front-wheel drive. In true American style, though, there's an impressive 230lbs/ft of torque, so 'effortless' is the best description as to how the engine behaves.

Features of the engine include an alloy block but cast iron heads, pushrod valvegear, and electronic sequential port injection. It is also placed transversely atop a four-speed automatic 'box with electronically controlled change points.

Suspension is by MacPherson struts all round with coilsprings at the front and a composite transverse leaf at the back. There are disc brakes at each corner along with Bosch's latest ABS III anti-lock system, and the 15ins alloy wheels are shod with tyres exclusive to the car, Goodyear Eagle 225/60VR15s.

The Allante is also a moving electronic marvel, with a computer whose memory has a capacity of 90.7K bytes (which means *enormous*). This has allowed the use of multiplexing which, by using digitally coded pulses, allows a single wire to do the job of many.

But the Allante isn't really about mechanicals. It's about look and image, and the Pininfarina connection. It's considerably more than just a badge job, too, such as the Ford-Ghia tie-up. Pininfarina actually make the body as well, and the assembly of the car involves the use of a fleet of Boeing 747 cargo aircraft ... The base chassis – floor pan and front and rear under-bodies – are flown out via said Jumbos to Italy where the Italian company add the body and interior, the assembly then being returned to Detroit where engine and suspension are grafted on.

It's an expensive and complicated way of doing things, perhaps, but Pininfarina can cope nicely with the production of the 7500-odd bodies envisaged for the Allante, a figure that would cause headaches in America where such a tiny run is regarded as very small beer indeed. One spin-off from the Pininfarina connection is that we should see the Allante being sold in Europe before very long, although with a $50,000 price tag in its home country it's going to be excessively expensive this side of the Atlantic.

Driving impressions

It was appropriate, somehow, that my first sight of the Allante should be outside the Novi Hilton, a plush hotel in the American idiom to the west of Detroit. It looked right at home in such luxurious surroundings.

First impressions are that it is a lot bigger in the flesh than it looks in photographs, and curiously anonymous: it doesn't leap out at you as would, for example, a 1959 Eldorado. Pininfarina's lines are neat and tidy if a little bland, subdued even – but that is very much the latest American trend anyway. Long gone are the days of bulk and chrome ...

It's a big car for a two-seater, and this shows up in the cockpit, which is enormous, particularly between doors.

Initial impressions are that it is incredibly smooth, ultra-refined, in true Cadillac traditions. At 65mph on the freeway, the loudest noise is either the radio (backing up Cadillac's claim) even with the top down, or the slap-flap of those fat tyres across the breaks in the concrete surface. You hardly ever hear the engine except when you floor the throttle – and even then it's just a distant hum – while you would swear the 'box is a single-speed device, so smooth are the changes. Waft is what it does best.

The surprise comes when you find a corner or two. Yes, it does understeer if you push it, but those fat tyres again give it a fair amount of grip, while the performance (124mph top speed, 0-60mph in 9.5s) isn't enough to have power fighting traction. It doesn't roll all that much, and the steering was a real surprise. It's typically American in being very light, but it's also more direct than in days of yore, and has quite good in-built feel. It's unlikely to stay with a Merc or XJS cross-country, but it won't be left *that* far behind ...

The cockpit is amazing. The seats are by Recaro but aren't the hip-hugging devices we usually get: instead they're fairly broad and flat without much side support, but, with a vast range of adjustment, plus a steering wheel adjustable for rake and reach, a good driving position is easy to achieve. The instruments are all electronic with both digital and analogue displays, and at last someone has come up with glowing dials that can actually, and easily, be read in direct sunlight. But what really stuns is the number of buttons which spread themselves across the dashboard. I never did have a chance to count them all ...

On the whole, the new Allante represents quite a break for one of America's oldest firms, a sporty but not a sports car, exclusive once more, with that 'made in Europe' cachet. On its home ground, I can see it giving the Merc and XJS some trouble ...

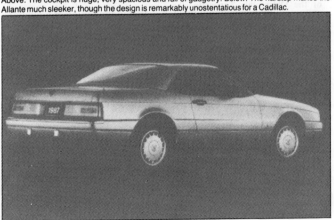

Above: The cockpit is huge, very spacious and *full* of gadgetry. Below: The hardtop makes the Allante much sleeker, though the design is remarkably unostentatious for a Cadillac.

ATLANTIC CROSSING

In a unique deal with Pininfarina, which involves flying bodies from Italy to America, GM has produced the Allanté. Mark Gillies looks at the luxury two-seater which, at around $50,000, should quench the American thirst for prestige

Digital dashboard *is complemented by extensive use of leather*

Once you cut through the morass of PR hype with which Cadillac has surrounded its new Allanté sports two-seater — part produced by Pininfarina in Turin — you find out that it's a pretty good motor car...

Starting point for this machine is foreign domination of the luxury sports-car market in the USA, and one car in particular: the Mercedes-Benz SL. Like Gucci loafers and Armani shirts, a Mercedes sports car is a designer label that tells everyone you have money. With capital letters.

Cars such as the Stuttgart product have shown GM that a status hungry and wealthy sector of the car-buying public is willing to pay over the odds for prestige — and to the chagrin of home automakers, that means buying foreign.

Enter the Italian coachbuilding concern of Pininfarina. The firm has had links with GM's luxury *marque* before, beginning in 1931 when a V16 Cadillac chassis was fitted with a glamorous body in Italy. The association was continued through the 1950s and '60s with the Cadillac Starlight in '59 and the Brougham coupé in '61.

But as well as concrete links. Pininfarina has held influence over GM designers. Present design director Chuck Jordan raves about the firm's products, and Bill Mitchell, once supremo at GM Design, said: "I visit Turin motor show every year, and to me, Pininfarina is a source of inspiration. After all, if I have to rob, I do it in a bank, not a grocery store."

Pininfarina also produces very good looking cars and can engineer body structures using fully integrated CAD/CAM systems. The name sounds good too: exotic, and with the *cachet* of being Ferrari's body stylist. Americans just *love* that kind of thing.

Cadillac went to Pininfarina four years ago with its LTS project (luxury two-seater), with the final agreement of production and design occurring in '83.

What ensued was a fairly odd agreement, and surely the most expensive way known to man of building a motor car. Pinin did all the body development work and styling, and carried out aero tests on site at its Grugliasco base and crash testing at Porsche's Weissach labs.

Andrea Pininfarina, in charge of the Italian end of the project, is especially proud of the wind tunnel work, which concentrated on promoting good airflow through to the radiator, reducing hood billow and wind noise buffet — Cadillac claims you can talk to your passenger at 65mph without disturbance. At that speed, Americans could theoretically have picked up a speeding ticket...

The car is part built in Turin. Pininfarina constructs the bodyshells, using robot technology for at least half the spot welding, and paints and trims them. Painting and assembly is carried out at the base plant in Grugliasco, with trimming at a new, purpose-built facility at San Giorgio, near Turin airport.

The cars are then 'crated' up for despatch to the USA, which is carried out by an airbridge operated by Lufthansa and Alitalia. There are three round trips per week, each carrying 56 bodies: GM claims the airbridge gives Cadillac 'the longest assembly line in the world' — is it also the most expensive? — and helps in quality control because it is faster than surface transport. After arrival in Detroit, the bodies are taken to Cadillac's sophisticated Hamtramck plant for mechanical components to be fitted.

The finished product of this trans-Atlantic crossing seems to have lost something in the translation: lovely looks, but a shame about the running gear, if you judge it from a European point of view.

For starters, it has front-wheel drive, which has never been an ingredient on the European luxury sports car menu. Reviving memories of the Oldsmobile Toronado, the engine for the fwd set-up is a 4.1-litre V8, which has an aluminium-alloy block with cast iron cylinder liners and cylinder heads. Cadillac emphasizes the high-tech features of the engine such as the sequential port fuel-injection system and tuned inlet tracts, but the low red line of 4750rpm gives away that this engine has a cam-in-block, pushrod overhead valve layout. A maximum power figure of 170bhp at 4300rpm isn't going to set the opposition quaking in its collective boots, but peak torque of 230lb ft at 3200rpm is useful. Cadillac emphasises that its new baby is no road-burning slingshot, but a car with usable performance — in other words, good old-fashioned, torque. ▶

One of Allanté's *best features is its shape. Clean and stylish, with the hood up*

V8 4.1-litre *produces 170bhp and has tremendous torque*

Easily attached *hard top is part of the Cadillac package*

or down, it is exactly right for the US market

Instruments *set new standards for interior display systems*

Allied to the transversely-mounted engine is a four-speed automatic 'box — no, you can't have a stick shift — which incorporates a viscous converter clutch and electronic shift control.

Why front-drive? Cadillac maintains that 'all weather traction and maximum space efficiency' are the reasons, but that sounds pretty tame. Why do you need space efficiency in a two-seater, for instance? Cadillac claims that boot space is enhanced, but any car which has only two seats is wasting space anyway. Then again, GM would never say that the drivetrain was adopted just because it happened to be available, off the shelf, as used in the Eldorado, Seville and De Ville.

The suspension set-up won't have Jaguar and Mercedes designers rushing back to the drawing boards, either. It's all independent, by MacPherson struts and coil springs at the front, with trailing and lateral links as well as an anti-roll bar, while there are MacPherson struts at the back used in conjunction with a composite transverse leaf spring. Cadillac claims a skidpan figure for lateral acceleration of 0.82g for Allanté, whereas a test figure for the latest version of Mercedes' evergreen sports car, the 560SL, is 0.78g.

Brakes are Bosch anti-lock, with ventilated discs at the front, solid discs at the rear, and steering has power-assistance. Goodyear came up trumps on the tyre front, developing the Eagle VL covers specially for Allanté. They are claimed to give high-performance car cornering power allied to luxury car tyre ride.

The area of Allanté's specification which the engineers rave about, however, is the level of electronics used. The memory capacity of the combined computer systems is 90.7K bytes, while Allanté uses Multiplex wiring — the use of one wire to carry out the function of many — for the first time in the car industry. Useful attributes of these electronics include the lights being turned on automatically at dusk and off at dawn, while, if a light fails, the system allows for another to be turned on as a substitute.

Cadillac has yet to price Allanté, but the guess is that it will be around the $50,000 mark (£34,000): for that, there is only one option; a custom-style cellular telephone. Everything else, such as 10-way adjustable seats with memory, high quality Delco-Bose stereo system, air conditioning and cruise control comes as standard. There's even a retractable AM/FM/cellular phone aerial fitted.

Undoubtedly one of the best features about Allanté when you inspect it is the shape. It is no show-stopping supercar, but it's exactly right for the market at which it is aimed. It's clean, handsome as opposed to pretty, and very stylish. Whether top up or down, it looks good, but can't be regarded as a trendsetter.

Sergio Pininfarina maintains that the outward appearance of the car is vital, comparing it to the first time one sees a girl. "The first impressions," he says, "are very important. If the girl is nice, you ask if she is well educated, speaks languages, if she is rich. But if she is ugly, it takes time to find out these things." Fortunately for Allanté, it doesn't take a great deal of time to find out the car's character.

One notable aspect of the car's exterior is the lack of brightwork and ornamentation. The grille, for instance, is one of the few parts which is plated, and even that is simple yet redolent of the American *marque*. Badging is discreet, with a Cadillac emblem visible on wheel trims, grille and on the US-mandatory third brake light, itself neatly mounted on the boot lid. There are Pininfarina badges on each flank, just behind the front wheelarches.

On the nearside (UK) tail lights there is a Cadillac script, while on the offside appears the one reference to the car's name. Allanté comes as a soft-top car, but the hard top which is part of the deal is easily attached and very stylish. Drag coefficient of the car, by the way, is 0.34, low for an open-top sports car.

Like the exterior, the interior gains from being European. It is tasteful and neat, carried out with excellent attention to detail.

Seats are trimmed in leather — as is the transmission tunnel and the bottom of the door trims — with a choice of two colours of hide and carpet. The seats themselves, courtesy Recaro in Italy, are very comfortable and supportive into the bargain, and have 10-way adjustment with two-position memory; switches are built into the left-hand armrest, along with window pulls and central door latch. Carpeting is lush and comprehensive.

Ahead of the driver is a bank of switches and dials, which set new standards for interior display systems. The two-spoke leather-bound steering wheel with mandatory cruise control button built into left-hand spoke is adjustable for rake/reach, and it shields a stalk which operates high beam and indicators. To the left of the wheel on the padded plastic facia is an air vent and a panel which contains touch buttons and dimmers for the lights. That works well, despite initial unfamiliarity, but the plastic facia, column stalk and bonnet latch would look more at home in a Chevrolet Cavalier than in the most expensive car on GM's inventory.

To the right of the wheel is another touch button panel, this time containing more cruise control switches and mirror and wiper adjustment, with the LCD instrument display above. Even the rev counter and speedo are LCD displays, but with traditional analogue dial shape, presumably so that traditional Cadillac customers aren't overcome by shock . . .

The use of LCD instruments is very clever, and I have to admit that I liked it — excepting the speedo, for which the digital readout seemed far more apposite than the 'fake' analogue readings.

The centre console, too, showed signs of Italian flair, with the whole assembly canted towards the driver,

Centre console *houses push buttons for stereo/heater controls*

Semi-powered *hood arrangement was*

and containing, from top to bottom, air vent, automatic climate-control-system which bamboozled me as soon as I got near it, a driver information centre, that immensely flashy hi-fi and a *pininfarina* script below. The driver information centre provides complete operating system diagnostics, and these, together with the secondary instruments, can be shut off at the touch of a button, leaving only revs and speed on the main instrument cluster.

The transmission tunnel continues in equally tasteful vein, with leather-trimmed gear selector and gaiter; there's a visual identification of gear selected on left-hand side of lever to complement the LCD indicator on the instrument display. There are two hinged covers, one lifting to reveal ashtray and cigar lighter, while the other has oddment space to store your *Hotel California* and *Hollywood Nights* tapes as well as the electric switches for boot and fuel filler latches.

Behind the seats is auxiliary luggage space which lives under the hood and is accessible from the boot, while behind the driver's seat are controls for the hood.

The bottom of these two unlatches the tonneau cover — itself a rather tacky plastic affair — to allow the hood to swivel up. In order to save weight, Pininfarina hasn't resorted to full electric operation — how will that go down with rich Americans used to their creature comforts? — so you fold down the glass back window, stow the tonneau cover back into place and operate the top catch, which in turn powers an electric motor to pull tight the back, part of the hood. The top of the hood is raised manually and held in place by two easily operated catches. One gripe: the pair of sunvisors, both of which incorporate mirrors — there's vanity for you — have to be swung clear to raise and lower the hood. The boot is spacious, beautifully trimmed and houses a space-saver tyre. Typical attention to detail, though, means that the cover over the tyre is extendable to allow standard size tyres to be stowed.

The Italian end of the equation, therefore, works admirably. Even with this writer's slight bias towards things Latin, it's fair to say it's an improvement in both interior and shape over Mercedes' now ageing SL. GM's customary standard equipment list, moreover, means that it's reasonable value.

The element that has been lost in the translation between Italian and American, though, is European driveability. For driving to the country clubs in the hills around Monterey or posing around Beverley Hills, the dynamics are fine. But for hustling along an Italian country road . . . discount it.

The major problem is understeer. *Terminal* understeer. Four litres of power, front-wheel drive and an auto 'box sound like a recipe for a disaster when the car's pushed, and if the predictable power-off tuck-in stops calamity, it still doesn't feel too good. However, brake early and drive through a corner rather than indulge in hot-hatch style hooliganism and the car is quicker and better than you'd imagined.

Lending itself to the feeling of luxury rather than litheness is the drivetrain. The V8 lump never becomes asthmatic because its red line's so low — but has tremendous torque: this car has shire horses under its bonnet, not New Forest ponies. So Cadillac has transferred emphasis from out-and-out acceleration to driveability: the 0-60mph time is quoted as but 9.5 secs, the top speed as 125mph.

The aspects which most impress are the ones which Europeans may regard as namby-pamby and dismiss as secondary in a sports car: there's the quietness at 60mph, which allows you to hear *Hotel California* properly; the slickness of the auto 'box, which changes under gentle driving as smoothly as a PR officer evades questions; excellent ABS brakes; a ride which soaks up bumps in its stride; and well-weighted power steering which lacks feel only above 70mph, by which time the CHiP bikers have clobbered you anyway . . .

Even the roll is so well damped for an American machine that seasickness tablets definitely weren't required.

I'm sure that if it came to a straight shoot-out between Mercedes, Jaguar and Allanté over country roads in Europe, the Allanté would lose: but in posing stakes it would run both close.

Allanté will almost certainly come to Europe, and maybe to Britain — although in that case, still as a left-hooker. Cadillac has produced some Allantés with conventional analogue dials, which it is claimed are part of the European package. Whether Europeans will take to the car is another matter, as is the pricing: but there's no doubt that in America, it'll deliver the goods . . . and frighten Mercedes. ■

SPECIFICATION

CADILLAC ALLANTE

ENGINE
Transverse, front, front-wheel drive. Head/block iron/al. alloy; 8 cylinders, 90 deg V, bored block, wet liners, 5 main bearings. Water cooled, electric fan.
Bore 88mm (3.46ins), **stroke** 84mm (3.31ins), **capacity** 4100cc (250 cu ins).
Valve gear ohv, 2 valves per cylinder, chain camshaft drive. **Compression ratio** 8.5 to 1. Breakerless electronic ignition, electronic sequential-port injection.
Max power 170bhp (PS-DIN) (127kW ISO) at 4300rpm. **Max torque** 230lb ft at 3200rpm.

TRANSMISSION
4-speed automatic.

Gear	Ratio	mph/1000rpm
4th	0.70	35.6
3rd	1.00	25.0
2nd	1.57	15.9
1st	2.92	8.6

Final drive: planetary, ratio 2.95

SUSPENSION
Front, independent, MacPherson strut, lateral and trailing links, coil springs, telescopic dampers, anti-roll bar.
Rear, independent, MacPherson strut, H-control arm, transverse leaf spring, telescopic dampers.

STEERING
Rack and pinion hydraulic power assistance. Steering wheel diameter 15.7ins, 2.97 turns lock-to-lock.

BRAKES
Dual circuits, split diagonally. **Front** 10.25ins (260mm) dia ventilated discs. **Rear** 10ins (254mm) dia discs. Hydraulic servo. Handbrake, foot lever acting on rear discs.

WHEELS
Al alloy, 7ins rims. Goodyear Eagle VL tyres size 225/60 VR15.

DIMENSIONS, WEIGHTS
Length 178.6ins (4537mm)
Width 73.4ins (1864mm)
Height 52.2ins (1325mm)
Wheelbase 99.4ins (2525mm)
Track F/R 60.5/60.5ins (1538/1538mm)
Weight 3493.8lb (1584.8kg)

PERFORMANCE (claimed)

	Auto
Top speed	124mph
0-60mph	9.5secs

EQUIPMENT
Automatic	●
Five-speed	N/A
Power steering	●
Electric windows	●
Central locking	●
Radio/cassette	●

● Standard N/A Not applicable

chosen to save weight. Sunvisors get in way of its operation

CADILLAC ALLANTÉ
TEES OFF AGAINST
MERCEDES-BENZ 560SL

Can Detroit really compete with Germany's legendary luxury convertible?

PHOTOS BY RICHARD CHENET

READERS UNDER 30 might find this hard to believe. But there was a time, not too many decades back, when Cadillac was the undisputed ne plus ultra of cars in America. The European car had been discovered already; GIs (that's World War II soldiers, whippersnapper) had brought MGs and Jaguars home from England, and a handful of enthusiasts, nourished spiritually by this magazine, were discovering the delights of Italian *Gran Turismo* coupes. As soon as the Germans could get their bombed-out factories working again, they began sending us their dignified, solid-as-bank-vault Mercedes and funny-looking Porsches.

For the first 20 years or so after the Big War, this was all lunatic-fringe stuff as far as most Americans were concerned. After all, who in their right mind would lay out $3500 for a little roadster with only six cylinders when you could have a real automobile—a full-size luxury car with a V-8 engine—for the same money?

That's how a comparison of the Jaguar XK-120 with a middling Cadillac sedan in the early Fifties might have looked to the average American motorist. And, to project the comparison a little closer to our times, who in his right mind would have chosen the $7054 Mercedes-Benz 280SL in 1969 over a $7110 Cadillac Fleetwood Brougham?

As time went on, a growing number, that's who. Inexorably, the word about European cars spread from New York westward, from California eastward: There was an alternative to the domestic automobile, which at the time was a rather uniform product. You got "less car" for your money: smaller body, fewer cylinders, less automation. But you also stood a chance of getting more: more roadability, sometimes more quality, and, admittedly the most subjective quality of all, better esthetics.

The Seventies put the whole movement over the top. Paradoxically—then again perhaps not paradoxically—that is when the price of fine imported cars skyrocketed. Freed in 1971 from fixed exchange rates, the dollar tumbled for the next decade; it took more and more of them to buy a given imported product. Imported luxury cars, which through the Sixties had cost about the same as domestic ones, soared past them on the price lists.

As a concrete example, consider Mercedes' SL. When introduced in late 1971 (as the 350SL 4.5), it cost $10,489. By 1981, with a smaller engine but more equipment, the price had ballooned to nearly $39,000. Today it is called the 560SL and it is enhanced by a larger engine, better automatic transmission and added standard equipment. It lands in U.S. showrooms at about $55,000 including its Gas Guzzler tax.

Normal laws of commerce would decree that a product whose price quintuples over a period of years when other prices were doubling to tripling just might suffer a decline in sales. Not the SL. From 6656 units in all of 1972 to 5987 in only the first half of 1986, the SL has experienced an almost steady climb in U.S. sales.

Today it is the darling of every country club, every doctor's wife and every struggling fashion designer. Short of the exotics, it has become the most desired car in the land. Who in Detroit-land, until much too recently, would have predicted that 10,000 Americans a year would be laying out nearly twice the price of the dearest U.S. car for a 2-place convertible without so much as power seats? And millions soundly rejecting Cadillacs as status symbols?

It happened, and some believe not in spite of the price climb, but at least in part because of it. Whatever the sociology, America's erstwhile prestige leader woke up in the early Eighties to find itself relegated to the upper middle class of automobiledom in its own land. Or, in some regions, even worse.

In 1982, Cadillac's planners belatedly turned some attention to the prestige market that had gone right past them—the lucrative ultra-luxury $50,000-and-up automobile market of Beverly Hills, Dallas and the Hamptons. In a most unusual move, General Motors not only planned a 2-place convertible but also let somebody other than GM Styling style it. Somebody European.

Allanté: America challenges European status

THUS HAVE Cadillac and Pininfarina labored mightily and brought forth a challenge to the vaunted Mercedes SL. According to the custom of our times, it carries a name—Allanté—conceived by one of those product-naming agencies for psychological effect.

Though it isn't literally a copy of the Mercedes SL, the Allanté was clearly designed to compete specifically with the SL. Their external dimensions are very close; their open 2-seater configuration with soft and hard tops are identical—the Cadillac, however, is substantially lighter.

So what we have is a Cadillac interpretation of the SL concept. But thanks to Pininfarina, the Allanté's style is as un-Cadillac as its underpinnings are Cadillac, and not at all an imitation of the SL's German-American look. Its lines are pure Italian, at once timeless and contemporary. And smooth. The Allanté's aerodynamic drag coefficient is 0.34 with the hard top on.

Cadillac's 1986 Eldorado/Seville supplied the underpinnings. Cadillac shortens the platform and ships it to Italy for its new skin; from there it is airfreighted back to Detroit for its innards. Wider wheels spread its tracks and carry specially developed Goodyear Eagle VL tires, which generate less rolling resistance and tread noise than Eagle GTs—and sacrifice some traction and cornering capability. Suspension springs, twin-tube shocks, bushings and quicker 15.5:1 steering are specific to the Allanté. The odd aluminum-block/iron-head 4.1-liter V-8 has been blessed with better breathing, port (instead of throttle-body) fuel injection and roller cam followers for a 40-horsepower boost to 170 bhp.

Otherwise, the Allanté works are pure Eldo. The V-8 sits crossways and powers the front wheels through GM's infamous Turbo Hydra-matic 440 4-speed automatic transmission, improved and renamed F7. Its front wheels are suspended on MacPherson struts, its rears on a unique strut system that uses a double-anchored transverse fiberglass rear spring as both spring and anti-roll bar. The four Eldorado disc brakes are controlled by the Bosch ABS III, the latest and most integrated anti-lock system available.

Allanté's cozy cockpit shows both Italian and Cadillac influences. We'd say the shape and layout of its instrument panel is more of the latter than the former. So were the 2-tone red-burgundy leathers and plastics of our pearlescent-white test car. But there's plenty of Europe in the modern, yet tasteful and restrained contours of dash and door panels.

At least as foreign as the Italianate body is the Allanté's price, not set at this writing but expected to fall between $45,000 and $50,000. That's about 70 percent beyond the fattest regular production Cadillac list price at the moment, uncharted territory for Cadillac—whose captains know as well as you and we know that when you reach these levels, price is no longer a mere matter of product value. It's a complex psychological matter, all wrapped up in status, perceptions of quality, expectations of resale value and Lord knows what else.

The Cadillac folk are going up against a product that's a sort of legend in its own time, commanding $55,000 without power seats or even two electrically adjustable mirrors. They know the SL didn't acquire this spell overnight. They understand that it did get there because it is good. A bit antiquated, yes, but built like a bank vault and finished with care unknown in American carbuilding.

No matter how good the Allanté turns out, it will take time—probably lots of it—for Cadillac to convince ultra-luxury buyers it is serious competition for Mercedes-Benz. The fact that Cadillac still builds De Villes, Fleetwood Broughams, Eldorados and Sevilles won't help.

PERFORMANCE

	Cadillac Allanté	Mercedes-Benz 560SL
Acceleration:		
Time to distance, sec:		
0–100 ft	3.3	4.0
0–500 ft	9.2	8.7
0–1320 ft (¼ mi)	17.1	15.8
Speed at end of ¼ mi, mph	80.0	91.0
Time to speed, sec:		
0–30 mph	3.0	2.8
0–60 mph	9.3	7.5
0–80 mph	17.1	12.5
0–100 mph	30.5	19.6
Top speed, mph	119	137
Test fuel economy, mpg	22.4	16.8
Brakes:		
Stopping distances from, ft:		
60 mph	168	142
80 mph	275	250
Overall brake rating	very good	excellent
Handling:		
Lateral acceleration, g	0.77	0.78
Slalom speed, mph	61.0	58.6
Interior noise with soft top/hard top, dBA:		
Idle in neutral	56/54	46/46
Maximum, 1st gear	69/70	79/78
Constant 30 mph	62/56	58/58
50 mph	66/64	66/62
70 mph	71/68	76/68

CALCULATED DATA

	Cadillac Allanté	Mercedes-Benz 560SL
Lb/bhp (test weight)	21.6	17.5
Bhp/liter	41.6	40.9
Engine revs @ 60 mph (top gear)	1600	2200
Brake swept area, sq in./ton	196	217

Mercedes-Benz 560SL: The established

Over its long career, the SL has had many different engines under its long, high hood: the original 3.5-liter V-8, which America never saw; the 4.5-liter V-8 that powered U.S. versions from 1971 through 1981; the aluminum 3.8-liter after that; and now a 5.5-liter monster V-8. For other markets, it has also been available all along with a 6-cylinder engine.

The present engine puts 227 bhp through a 4-speed automatic transmission to the rear wheels, which are suspended on vintage semi-trailing arms with extra linkage to control the massive torque. Classic unequal-length A-arms handle the front wheels, and since 1985 anti-lock braking has regulated the large 4-wheel disc brakes.

Minor styling changes and modification of the bumpers to "weak 5-mph" protection marked the 560SL; otherwise the SL's shape has been unchanged since 1974. Its drag coefficient is 0.43 with hard top, 0.41 with soft top. Inside, the 560SL is spartan, lacking many of the Sybaritic luxury and convenience features that have become *de rigueur* in top luxury and GT cars. But nearly everything there is of the best quality.

Allanté versus 560SL: Can Cadillac compete?

Against Mercedes' 100 years of building cars and 15 with the present SL, Cadillac brings 85 years of building cars and about four years of development time on the Allanté.

We must tell you right off: The Allanté we tested was not a production car, but something between a prototype and an early production unit. Production cars could be better. Or worse: Allanté engineers did not try to convince us production cars would be different, except in a few details.

That said, we plunge into the results of 2½ days of testing, driving, examining, probing and, finally, evaluating Cadillac's challenger and Mercedes' incumbent.

Powertrain and performance

THE ALLANTE'S 4.1-liter V-8 isn't cut out for high output. After it moved smartly off the line, it soon went flat. Not that the Allanté is deficient on performance—0–60 mph in 9.4 seconds is more than adequate for American driving, and Cadillac has fitted an exhaust system that adds auditory spice to every accelerator movement.

But the 560SL has it all: torque, power, revs (redline 6000 vs Allanté's 4700). All except fuel economy, that is: On our test trip the SL turned in only 16.8 vs the Allanté's impressive 22.4 mpg. So all that performance is not without penalty, but admonish as we might about petroleum being a finite resource, it's unlikely fuel economy will be much of a factor for buyers of these cars. Cruising range might be; the Mercedes will have to be fed every 360 miles or so, the Cadillac only every 470.

In the transmission department, too, Cadillac's dependence on existing GM componentry works against the Allanté. Though internal changes may have freed the 440 box from its reliability problems, other deficiencies remain. The engine could use shorter ratios. A typical American T-handle shifter and the transmission's slow response to manual changes frustrate any attempt to intervene manually. In its favor, Cadillac's unique viscous torque-converter clutch engages and disengages with far less frenzy than conventional ones.

By contrast, the SL is automatic-transmission nirvana. Its high internal-torque multiplication assists the already mighty engine, and Mercedes' uniquely gated shifter remains the model for all others. Being European, Mercedes also realizes that skilled drivers sometimes shift automatics manually; this one responds immediately. Bottom line: Though thirsty, the Mercedes wins here.

Chassis: handling, ride and braking

LEST WE FORGET, Mercedes also uses off-the-shelf components in its 2-seater, and in the SL's case the components date back to 1968 sedans. So this isn't Mercedes' best effort—the 190 and 300 series have far better suspension.

Thus the 560SL manages only a middling 0.78g on the skidpad; slow steering and what seems like too much lateral softness conspire to make it an unresponsive handler when you push it. On the skidpad the Allanté is about equal to the SL. But on the road and in the give and take of the slalom test, the Cadillac's quicker steering, wider tires and tighter suspension, both vertically and laterally, pay off in better reflexes.

With classic logic, the Allanté extracts its penalty in riding comfort. It's a little stiffer, a little jigglier than the soft-riding SL. And, despite Cadillac engineering's claim of greater body rigidity, a greater degree of shaking and flexing detracts further from perceived comfort.

Both cars have 4-wheel disc brakes and Bosch ABS; the Mercedes has larger brakes, the Cadillac a newer anti-lock system. In everyday driving we found the Cadillac's braking feel superior, but in emergency braking tests the Mercedes stopped dramatically shorter—well over a car length from both 60 and 80 mph—and with much better directional control.

GENERAL DATA

	Cadillac Allanté	Mercedes-Benz 560SL
Base price	est $48,000 (FOB Detroit)	$54,600 (east & Gulf)[1] $55,030 (west)[1]
Price as tested	est $48,000	$55,020 (east)[2]
Layout	trans front engine/fwd	front engine/rwd
Engine type	ohv V-8	sohc V-8
Bore x stroke, mm	88.0 x 84.0	96.5 x 94.8
Displacement, cc	4087	5547
Compression ratio	8.5:1	9.0:1
Bhp @ rpm, SAE net	170 @ 4300	227 @ 4750
Torque @ rpm, lb-ft	230 @ 3200	279 @ 2750
Fuel injection	GM electronic	Bosch KE-Jetronic
Fuel requirement	unleaded premium, 91 pump octane	unleaded premium, 91 pump octane
Transmission	4-sp automatic	4-sp automatic
Gear ratios: 4th	0.70:1	1.00:1
3rd	1.00:1	1.44:1
2nd	1.57:1	2.41:1
1st	2.92:1	3.68:1
1st	2.92 x 1.6:1	3.68 x 1.9:1
Final drive ratio	2.95:1	2.47:1
Steering type	rack & pinion, power assisted	recirculating ball, power assisted
Brake system, f/r	10.2-in. vented discs/ 10.0-in. discs; hydraulic assist; ABS	10.9-in. vented discs/ 11.0-in. discs; vacuum assist; ABS
Wheels	forged alloy, 15 x 7JJ	cast alloy, 15 x 7J
Tires	Goodyear Eagle VL, 225/60VR-15	Michelin XVS, 205/65VR-15
Suspension, f/r	MacPherson struts, lower lateral arms, trailing links, coil springs, tube shocks, anti-roll bar/Chapman struts, lower lateral A-arms, transverse fiberglass leaf spring, tube shocks	unequal-length A-arms, coil springs, tube shocks, anti-roll bar/semi-trailing arms, coil springs, tube shocks, anti-roll bar, torque compensator

[1] 1986 560SL price, includes $1300 Gas Guzzler tax; [2] includes heated seats ($420).

Cadillac Allanté 4.1-liter ohv V-8 sits sideways, drives front wheels.

Mercedes-Benz 5.6-liter sohc V-8 drives rear wheels in classic fashion.

Here, as it's ABS-cycled, the Allanté slewed to and fro in a sort of slow-motion dance. And, though equipment problems prevented us from measuring brake fade, an earlier test of the 560SL revealed no fade in our usual 6-stop test; we did perceive some fade in an uninstrumented evaluation of the Allanté's behavior in repeated stops. Bottom line: Cadillac wins on handling, Mercedes on ride and braking.

Interior: ergonomics, luxury and four tops

MERCEDES MUST be credited with developing the concept of the luxurious open 2-seater, though Ford originated it with the Thunderbird in 1955. The present SL has gradually become more luxurious since its 1971 introduction, acquiring more standard features like electric window lifts, central locking and automatic climate control as well as a bit of wood trim.

But Cadillac has leapfrogged the SL in this respect. Just the appearance of its interior is far more lavish than that of the SL cockpit, even though there's no wood and the plastics used seem a trifle less fine than those of the Mercedes.

But it's in luxury features that the Allanté tops the SL most dramatically. Where the SL has mere mechanically adjustable seats with a rudimentary height adjustment for the driver, the Allanté boasts firm, supportive Recaros with no fewer than six 2-way adjustments for both seats: fore-aft, backrest angle, height, separate front-of-cushion/back-of-cushion height and a highly effective lumbar-support variation. The six rocker switches aren't ergonomic heaven, but they work. Cadillac includes a tilt-telescope steering wheel too, so the Allanté accommodates various statures superbly.

In addition, the Allanté seats have two memory captures for the doctor and his wife, and an exit feature we don't recall seeing elsewhere: For easy exit and entry, touch that button and the seat motors to its rearmost, lowest position. Entry and exit can be further aided by tilting the wheel up.

This defense of the SL's seats: Simple as they are, they still provide adequate support and lateral grip and are easy enough to get into and out of.

Other luxury or convenience items Cadillac has over Mercedes include remote fuel-filler and trunk openers (not very convenient, though, because they're in the center glovebox for security), a trip

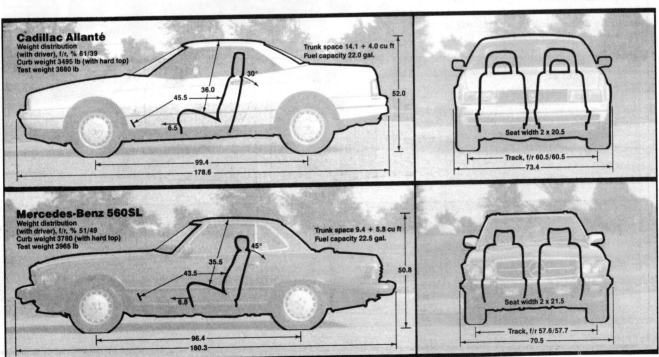

computer, dual electrically adjustable outside mirrors (Mercedes: right-hand only), a heated rear window in the soft-top as well as the hardtop, automatic headlight operation, headlight washers and the option of a built-in cellular telephone. The SL has three items the Allanté doesn't: heated seats, a central locking system (Allanté: electric door locks only) and a supplemental restraint system with driver airbag and knee bolster plus automatic tensioning seatbelts.

The Cadillac soft-top's glass rear window avoids the SL plastic window's inevitable creases and scratches, and Cadillac has provided rear quarter windows in the soft-top and hardtop. But the plastic top boot, an unattractive piece of polyurethane, conspires with the high rear deck to restrict rearward vision severely when parking. Views are better from the SL.

Cadillac's LCD graphic speedometer and tachometer are laid out in classic round black-on-white style, but instead of pointers they have bands of little yellow diodes crawling around the scales. There's a digital speedometer readout too. To the right are coolant, oil-pressure and fuel gauges in straight-line graphics; between speedo and tach is a colorful display of warning lights.

Except that we'd rather see needles—easier to read than the edge of a band—the speedo and tach show that electronic graphics are getting there. There is some washout of the graphic readouts in bright sunlight, something that's likely with the top down. A year or so after introduction, conventional dials will be an option.

Pushbuttons control almost everything in the Allanté. For the climate system, they're logical and satisfactory; for the stereo we'd prefer more tactile controls for tuning and volume. Three elements of the Allanté's controls struck us as particularly ill-advised: Wiper-washer controls are on the dash instead of the steering col-

CADDIE AND PININ

As Foster Brooks would say, Cadillac and Pininfarina go W-a-A-A-Y back. They had their first date in 1931; Caddie wore a V-16 engine on her long chassis and Pinin Farina (the way he signed it then) brought a snazzy 4-door speedster outfit, with a cover over the back seat instead of a raccoon coat. Everyone who was anyone thought they were all the rage.

But somehow they didn't really hit it off. Separated by an ocean and then by WWII, they didn't get together again until a couple of times in the Fifties and Sixties, Caddie in her shorter V-8 chassis and Farina in different 2-door sweaters and shirts but always with straight, nicely creased slacks. They weren't quite the sensation this time, even though most people thought they looked pretty good together. They went their separate ways again, Caddie being pretty much of a home girl at heart while Pininfarina played the field.

Then in 1982, by which time they were in their middle years but still far from over the hill, someone said to Caddie, "Remember that Pininfarina fellow? So charming, such a good dresser. Why don't you look him up?"

"Oh, I couldn't," Caddie said. "I wouldn't dare! Unless, maybe . . . Do you think . . . Oh, yes! Yes!"

Well! If you've been to *any* parties, seen *any* shows, read *any* magazines, you know

They Are An Item Again, going to *all* the country clubs and shops and theater openings together. Caddie has her tight, front-drive outfit and Pininfarina brings his trim, bare-headed or top-hatted, combination-sports-or-evening clothes. Caddie and Pinin look really terrific this time, so much younger than we remember them, and only that hussy Mercedes (you know all about *her*) could deny it!

—*Your Society Scribe, Jonathan Thompson*

1931.

1953.

1958.

1959.

1961.

umn, the headlight controls are next to unintelligible and the transmission selector's console indicator is obscure at best (it's repeated in the tachometer face, but the natural tendency is to look toward the console).

It'll probably be another decade before you see electronic readouts in a Mercedes. Well and good. They may not be trendy, but the SL's mechanical analog readouts—same information as in the Allanté—do the job and we rated them better.

We tested the cars in hot, muggy weather and thus can't comment on the heating side of their automatic climate control systems. True to folklore, the American air-conditioning performance is better than the German; mainly, it delivers more air with no more noise and cools the interior more quickly.

The SL has long been the world standard for convertible convertibles, sporting a reasonably easy-to-fold, easy-to-erect soft top and a hard top that turns it into the moral equivalent of a closed coupe. Cadillac wanted to do at least as well.

With both cars, it's best to read the owner's manual first before trying anything—Mercedes even provides a separate booklet on the tops. Once you've done that, managing the soft top is approximately on a par. The same comment applies to the hard top, but here the Cadillac has an edge because its aluminum hard top weighs only 58 lb—Mercedes' weighs 92.

The payoff is that with its hard top on, the SL is as quiet and virtually as solid as a closed Mercedes—and that's impressive, indeed. Despite claims, the Allanté is perceivably less rigid and rattle-free. Wind noise, which the Mercedes' hard top beautifully seals out, is flatly unacceptable in the Allanté even if the total noise level is little different from the SL's. Cadillac promises it'll be better when production starts, but doesn't promise the problem will truly be solved.

With soft tops up, the comparison changes; here the Mercedes is noisier at cruising speeds, not from leaks around the windows but because wind passing over the fabric roars so loudly. With the tops down, Cadillac wins by a noticeable, if not dramatic margin; fixed wind wings on the doors appear to help.

You don't buy 2-seaters for practicality, but the Allanté does have more cargo space and a fold-down partition connecting trunk and interior. Cadillac also lines the trunk lid with carpet; Mercedes leaves it unlined. Give Mercedes credit for a full-use spare tire against Cadillac's limited-use compact spare, but Cadillac has made a clever provision for temporarily carrying the full-size wheel you pulled off. Bottom line: Cadillac wins on luxury, but Mercedes has the fine touch of quality.

Conclusion

PICKING a clear winner in this upstart-versus-incumbent test wasn't possible. We confess a certain predilection for the new and fresh, which favors the Cadillac. But the Mercedes is remarkably good for a car having its basic engineering done in the Sixties.

Legendary quality is Mercedes' primary stock in trade, but brilliant performance and outstanding ABS braking have freshened it this year. Against these attributes, Cadillac brings better handling and greater luxury to bear while failing to match Mercedes' performance and quality.

As our ratings and staff preferences show (see table), the Allanté doesn't quite match the Mercedes objectively; in more subjective areas, we found enough to bring it abreast of the SL—our own subjective judgment, of course. But wait: abreast of the *current* 560SL. Precisely because that is an old concept, Mercedes is at this very moment preparing its successor. See the accompanying piece in "Miscellaneous Ramblings" about the new SL.

For 1987, though, it's new Allanté versus old SL. And as you see, Cadillac's close. It's too early to predict how it'll fare in Beverly Hills and the Hamptons, but we have it on good authority that JR will be driving one real soon.

CUMULATIVE RATINGS—SUBJECTIVE EVALUATION

	Cadillac Allanté	Mercedes-Benz 560SL	Comments
Performance:			
Engine	7.5	9.0	Cadillac is more fuel-efficient, Mercedes more powerful; both sound terrific.
Transmission	7.0	10.0	Cadillac is smoother, but Mercedes responds faster and has better selector.
Steering	8.5	7.5	Both have good road feel, but Cadillac's is quicker and more responsive.
Brakes	7.0	9.0	Mercedes' stopping distances are shorter and directional control is better.
Ride	7.5	9.5	Cadillac's ride is tighter, noisier; Mercedes' is more comfortable.
Handling	9.0	8.0	Though slightly slower on skidpad, Cadillac handles more responsively.
Body structure, hard top on/off	8/6	9/7	The SL remains the world standard for open cars; with hard top on, amazing.
Average	7.6	8.7	
Comfort/controls:			
Driving position	9.0	7.5	Allanté's power seats and steering-column adjustments accommodate everyone.
Controls	7.0	8.0	Some confusion in both; Mercedes wins mainly on wiper-washer control.
Instrumentation	8.0	8.5	Cadillac's electronic graphics are best yet, but tradition still triumphs.
Outward vision	6.5	9.5	Allanté's high rear end and soft top restrict vision for backing up.
Quietness	5.5	7.5	Rattles and wind noise (even with hard top on) detract from the Allanté.
Heat/vent/air conditioning	10.0	8.0	Cadillac wins on air delivery and easier-to-understand controls.
Ingress/egress	8.5	8.0	Allanté has tilt steering wheel and power seats with exit feature.
Seats	8.5	7.0	SL has good traditional seats, but Cadillac's 12-way Recaros steal the show.
Luggage space & loading	9.0	7.0	Allanté has more space, better finish and pass-through to interior.
Average	8.0	7.9	
Design/styling:			
Exterior styling	9.0	5.5	Restrained, yet contemporary Allanté styling emphasizes SL's age.
Exterior finish	7.5	10.0	Allanté suffers detail glitches and horrible top cover; SL is nearly perfect.
Interior styling	9.0	6.5	Allanté interior is a sportier, more exciting place to be.
Interior finish	8.5	9.0	Cadillac's materials, though attractive, are a bit less rich than Mercedes'.
Average	8.5	7.8	
Overall average	7.3	7.3	
Staff preferences:[1]			
Price independent	3	3	
Price dependent	3	3	

[1] With preferences, 1st choice = 2 points, 2nd choice = 1 point. Two staff members involved, voting 2-1 and 1-2, respectively.

ITALIAN BLOODLINE

THE LATEST AMERICAN SPORTING CADILLAC HAS AN INTERESTING HERITAGE, AS STUART BLADON EXPLAINS. IT IS ALSO THE FIRST CAR WHERE PRODUCTION REALLY DOES FLY . . .

When the Americans set out to do something, they certainly tackle it with verve, originality and ingenuity; and they have the national advantage that when they encounter a problem, they can throw money at it. All these ingredients are to be found in the project for the new Cadillac Allanté, a car that at first glance perhaps doesn't deserve space in our pages. It does, however, have a sporting heritage and a surprising heritage at that.

General Motors wanted a top rank sports car for its Cadillac division, and decided to seek European styling for the car, once again going to the Italian specialist Pininfarina of Turin for the body design. It was the latest in a long line of American/Italian cooperation, but this time it was different.

"When I received the Cadillac representatives, offering us cooperation in the LTS programme," said Sergio Pininfarina, "I could not believe my ears!" Not only would Pininfarina design the car, but they would also build the bodies for it.

Those preliminary discussions were held early in 1982, and in October the same year, the styling was approved in Detroit. It had to be readily identifiable as a Cadillac; but otherwise Pininfarina was allowed complete freedom to design the Allanté as he wanted. April 1983 saw completion of full-size models, and in the following September the contract to build 8000 bodies a year for six years was signed in London. If that was rapid by motor industry standards, the next step was light years ahead. The next step was: the air bridge. A development that proudly claims to give the Allanté the world's longest production line.

Bodies are to be fully painted and trimmed, with wiring, glass and all finish details completed before despatch to America. The prospect of such fine work being ruined in transit, if not by dockers with their well-known habit of walking over tight-packed cars in the hold of a ship, then by exposure to several weeks in a highly corrosive atmosphere, was not to be contemplated.

The imaginative alternative adopted was to fly the bodies direct from Turin to Detroit, 56 at a time, by Boeing 747 Jumbo. Special light-alloy racks have been built which carry the bodies one above the other across the width of the Jumbo, and it is anticipated that transit damage will be contained to the absolute minimum. The bodies go straight on to the racks at the factory and are not dismounted until arrival in Detroit.

The roots of this remarkable collaboration with Cadillac goes back a long way. In 1931, Pinin (or 'Tiny') Farina, father of the present chairman, Sergio, built a spider version of the V16 Cadillac for the Indian Maharaja of Orchha, for use on tiger hunting expeditions. It was re-imported to the US after the war, completely restored, and sold recently at an auction, fetching half a million dollars! It was one of the first cars that Pinin had built, and certainly the most famous.

The first major building contract came in 1959, when Pininfarina was called on to produce 2000 Cadillac Brougham saloon bodies; and later, many GM-designed prototypes were built by Pininfarina. The energy crisis temporarily interrupted the liaison with Cadillac, and several programmes for co-operation were discussed but never came to fruition – until Allanté

Pininfarina's facilities at Cambiano and Grugliasco could not cope with the output required for Allanté, running at nearly 700 a month, so in record time, a completely new plant was built near Turin at San Giorgio, and it was there that we went to see the body production slowly beginning to get under way and to drive an early example.

It has much of the typical American car about it, with the inevitable feeling of great bulk and width that is rather an embarrassment on the narrow Italian secondary roads; but it is not a sloppy car. The steering lacks the precision that you would expect in a thoroughbred, but nor is it vague and woolly in the American style such as to make you wonder just how much rubber is used in the linkage.

Reasonably taut, the suspension gives a good ride, and the car has a well-balanced feel on corners, enabling it to be hustled through without suffering great armfuls of understeer. It remains flat and level on tight corners and does not squeal its tyres unduly easily. Especially creditable is the way in which you are not really aware of which end puts the power on the road. It is, of course, a front-drive car, with its big V8 all-alloy 4.1 litre engine neatly fitted in transversely in the manner favoured by GM for most of its cars.

Cadillac engineers explain that they were anxious not to spoil the Allanté's reputation by incurring the US 'gas guzzler' tax. The V8 engine has unusually shaped pistons, cut away on each side to minimise friction, and fuel is injected into the ports sequentially in relation to the intake stroke for each cylinder. But the engine is evidently throttled back fairly heavily on port design, and produces only 170bhp – not much for 4.1 litres, and not much, either, for a car weighing 3500lb.

Standard transmission is a GM four-speed automatic transaxle, with four ratios and provision of a viscous clutch to restrict torque converter slip. The transmission has a very responsive change-down when extra acceleration or hill climbing is called for, but no attempt has been made to give the transmission a sporty selector lever.

The American driver, no doubt, will be happy to shift into position four and leave it there; but at one section of our drive a long hill with many tight bends had the transmission surging up and down between third and fourth ratios. A much safer, smoother and more sporting style of driving was achieved by manually selecting second and third as required.

In two important areas, the Pininfarina designers have excelled. The first is the superb choice of body shape and windscreen angle to ensure that fast driving with the top removed is a pleasure, without any buffeting or the feel that the slipstream is attempting to tear the occupants out of the car. Up to about 80mph, the wind noise and disturbance are impressively low, so that conversation can continue without any need to shout.

The other pinnacle of achievement is the ease with which the Allanté is converted from closed to open, and back again. The all-too-short test route started with a short *autostrada* section, and then there was a halt where, we had been forewarned, the hardtop, which is a standard fitting with every car, would be removed. I thought that a picture of that process would be a good idea, but by the time I had taken the camera out of its bag, the hardtop was off and being stowed in a roadside pallet to await our return. It is made of aluminium, and light enough for two people to handle easily.

The hardtop snaps down with an ingenious fastening system on to a rigid tonneau cover which slides rearward to reveal the folding hood, again a standard feature. It has to be pulled up manually, and there are two well-designed over-centre fastenings on to the sides of the windscreen header rail. Then the tonneau panel slides back in and the hood tensions down on to it automatically. The car also scores top marks for the rigidity of the body and lack of scuttle shake on bumps.

The Cadillac/Pininfarina connection goes right back to 1931

Advanced electronics and computer technology are exploited in the Allanté which, like the new Jaguar XJ6, uses the technique of single wire switching by coded signals instead of taking all wires up to the switches and back to the relative lamp or component. An ingenious function of the computer is that it will detect failure of any bulb, and where appropriate it will command another nearby light to come on. For example, in event of failure of a dipped headlamp, the adjacent fog lamp would light up.

The fasia takes on almost the appearance of the cockpit of a light aircraft, with all switches and minor controls in banks angled towards the driver; and a trip computer giving the usual readings such as fuel consumption and average speed is made to seem spectacularly complex. Centrepiece of the display is a self-illuminated instrument display using a form of gas plasma speedometer and rev counter to give analogue indication of speed and revs, plus a large and clear digital read-out.

One could never see the Cadillac Allanté as a serious competitor to such cars as the Jaguar XJSC, and its US price is expected to be in the region of 50,000 dollars. But in relation to American road conditions with their very low speeds, and their desire for a car that is extrovert, big, light to drive, and convenient to use in open top form, the Allanté has a lot to commend it, and the Jumbo shuttle service should be kept busy for a long time to come. ∎

Cadillac Allante

photography by David Fetherston

Daydream believer

Cadillac scrubs up its image with a sports coupé to take on the Europeans

by David Fetherston

Something wasn't right — the steering wheel hub stated 'Cadillac', but I was driving a sports car. Was I daydreaming? No, I was driving a revolutionary new Cadillac sports coupé, the $US60,000 Allante.

It took a few days to get used to it, but all my ideas about the kind of car Cadillac represented went sailing out the window. Initially I supposed that although this new car would be a reasonable automobile, it would not be a true sporting coupé. I was wrong. The Allante is a world class sports coupé and a landmark vehicle for the American auto industry.

The Allante was designed by Pininfarina on a commission from Cadillac, part of the deal stating that the bodies were to be built in Italy. The Allante's structure is based on a Cadillac Eldorado platform. These are air freighted to Italy where they are shortened by 8½ inches (216 mm) and trimmed up front and back. The body is then built up on Pininfarina's production line. They are also painted at this stage, and the interior and dash, air conditioning and steering column fitted. When the bodies are turned out they are flown back to the GM Detroit Hamtranck Assembly Center, where the powertrain, some electronics and suspension are installed.

What's it like to drive? Performance is nippy, and the Allante is easy to drive at speed even on twisty back-country roads. On the freeway it will haul along way over the limit while you carry on a normal conversation even with the top down. Top speed is quoted at 125 mph (201 km/h), and I found the 0 to 60 mph (96.5 km/h) time was right on the money, at 10 seconds. For a 3500 pound (1600 kg) car it gives reasonable zip off the line. Its mid-range acceleration is even better, with brisk performance in the 40 to 80 mph (65 to 130 km/h) zone.

The Allante's front-wheel drive assembly uses an aluminium blocked, wet sleeved 4.1-litre V8 fitted with cast iron heads, increased by 50 bhp (37 kW) from its stock version in the Eldorado/Seville to 170 bhp (127 kW) and 313 Nm of torque. Its redevelopment for the Allante includes electronic port injection, 8.5 to 1 low-friction pistons, roller cam followers and tubular stainless steel exhausts headers. The fuel injection uses a very stylish set of tuned length runners curving back 90 degrees to form an interesting centre package for the engine compartment. The underhood engine noise is reminiscent of a Mercedes or BMW mechanical purr — a pleasant change from the traditional clank of American V8s.

The engine sits sideways, driving a four speed automatic lock-up transmission. Power is transferred to the gearbox with a double roller chain from the output shaft off the torque converter.

Conversion between hardtop, convertible and softtop isn't quick, but it's pretty straightforward. Seating is great; instrumentation is perhaps a little gimmicky.

The F7 transmission is a heavy duty version of the THM440 and is beautifully smooth: under hard acceleration it will produce good clean shifts without coarse, hard kickdowns or ups.

On the road the combination of the Allante's weight, power and handling give it a firm and solid feel. Even with the complete powertrain up front, the car does not have a nose-heavy feel or attitude, and under heavy steering input the fully independent suspension is quite capable of keeping all the wheels in the right place at the right time. The front suspension uses MacPherson struts, loaded with coil springs, lateral and trailing links and a sway bar. The rear also uses MacPherson struts but with an interesting twist: the struts are located with a set of lower A-arms but are sprung with a transverse plastic leaf spring. It all rides on VL 225/60-15 Goodyear Eagles developed especially for the Allante. These are mounted on special alloy rims and give a great bite on any surface.

Pressed hard, the suspension works admirably for a car that has a 61 front/39 rear weight bias — the Allante does not feel like a heavyweight front-drive car. The nose points where you want and the tail follows. There is minimal ploughing and it gives good turn in, which is really quite surprising in view of its bias. Its handling is precise, manoeuvrable and it rides like a champ. The steering is power rack and pinion and has a 38 foot (11.58 m) turning circle.

The brakes are also of the first order. The Allante uses a Bosch ABS III electronic system with hydraulic assist hooked up to 10.2 inch (259 mm) vented fronts and 10 inch (254 mm) solid rear discs. These haul the Allante down to a halt with great ease, clattering quietly if pressed hard.

The interior space is great for two. Both positions have 10-way power adjusters on the Recaro leather-clad luxo seats. The view forward is fine, but with the low positioning of the seat base and the height of the headrest the view to the rear is somewhat diminished, especially with the hard top in place.

The Allante is a hardtop, a soft top and a convertible and works equally well in all modes. The conversion from hardtop to convertible and back to soft top takes a little time, but once learnt is relatively simple. The soft top looks as good as the hard top, and in convertible mode the interior is amazingly quiet at speed considering the open space surrounding you.

Turn the engine on and the blank black dash lights up like a fun park, with a mix of colourful graphic and digital images. Two large circular bar graphs take the place of standard tachometer and speedo. Both of these are difficult to flash read, but Cadillac claims they allow the driver to sense, rather than read, the tachometer and the speedo. Luckily the speedo is backed up with a readable digital readout, as I do prefer knowing exactly what the instruments say rather than having to 'sense' what might be occurring.

One gripe is the lack of space to store small objects. There is a bin under the centre console, but you have to first lift up the large armrest and then open the lid to get to the bin. I would have preferred several large pockets built into the door panels as extra space.

Cadillac did not intend the Allante to be a world-beating sports car. It is designed for a specific market, mainly those who have owned Mercedes before and want a blend of luxury, quality and exclusivity. This is a sector not tapped before by an American manufacturer, and so far it has been a success — the first 4000 production run has sold exceptionally well. Here is a vehicle that will sell to both the sporty driver and the cruiser who wants style, performance and luxury bound together in an exotic automobile. □

Style, performance and luxury — a very attractive combination.

BUICK REATTA AND CADILLAC ALLANTE

Giving luxury car buyers what they need isn't the same as giving them what they want

COMPARISON R&T ROAD TEST

MONEY ISN'T EVERYTHING*, as they sang on Broadway in the Big Musical Era, and even though the lyrics went on to put cash well ahead of whatever was in second place, the sentiment is and was correct: Sometimes you get what you pay for, and sometimes you don't care what you pay.

This is just as well in the present case because what we have here are two personal cars. When people use terms like "personal car," they hint, accurately, that we are paying for more than mere transportation.

One of these cars, the Buick Reatta, is an expensive 2-only-

*From *Allegro* by Rodgers and Hammerstein, copyright 1947.

seater. The other, the Cadillac Allanté, has the same number of seats, two additional cylinders and costs roughly twice as much. Back in November 1986, when the Allanté was introduced, we examined how it stacked up against that prototypical high-buck 2-seater, the Mercedes-Benz 560SL. Now it's time to see where Buick's entry fits into the picture.

Cadillac and Buick were assigned/allowed the project of creating cars to suit this class. Each division's designers were given some freedom and held to some limits.

A major limit begins with the basic configuration.

Chuck Jordan, GM's vice president for design, doesn't criticize his company. Instead, he concedes that varieties of mass

PHOTOS BY DAVID W. BIRD II

PHOTO BY RON SESSIONS

confusion sweep periodically across the world's drafting tables and that one such collective hysteria was the notion that all cars, of all sizes and classes, had to be front drive.

This error was at its most virulent in the U.S. It has been cured, Jordan says; witness GM's decision to continue the Camaro and Firebird as rear drivers and to retain and improve the full-size big sedans in their classic configuration.

But the return of common sense came too late—our interpretation, Jordan didn't say this—to help the Allanté and Reatta.

Another limitation common to both is GM's dictatorship of the accounting department. For reasons of presumed economy, the new cars were required to be based, literally, on existing pans, those used by the sibling Cadillac Eldorado and Buick Riviera. Thus the two cars were going to be similar, cousins in a way, with a few shared parts in the trimmed-down pan and some of the rear suspension.

Cadillac and Buick teams didn't have clean sheets of paper. But GM has saved money.

Enough complaints. Once they had their parameters, the Cadillac and Buick teams went off in separate directions.

The Allanté is a convertible, with a removable hard top, a folding soft top and only two seats because the marketing department believes that's one way those who don't worry about being practical can let people know it.

The engine is a 4.1-liter V-8, to be enlarged in 1989 to 4.5 liters as with other Cadillacs, with fuel injection and rated at 170 bhp. The only gearbox is a 4-speed automatic.

The body was designed by Pininfarina, assembled and painted by this company at a plant in Turin, Italy and shipped to Hamtramck, Michigan where the drivetrain is added.

The accountants must have lost this one to the marketing department. Odd thinking, as one must wonder: If they—Cadillac execs, that is—figure Americans can't design and build good cars, why should we, the buying public, think Americans—er, Cadillac can? No doubt we aren't supposed to notice.

Extra points to Buick, then, because the Reatta was designed at Buick headquarters in Flint, Michigan and the GM Advanced Studio No. 2, Warren, Michigan. And it's built at the Reatta Craft Centre in Lansing, Michigan by teams of workers trained to work as teams and to have responsibility; speaking of borrowing good ideas when you find them, we'll let you guess which direction that came from.

The Reatta is powered by a new version of the GM V-6, displacing 3.8 liters. It's new in that the V-6 now has a counter-rotating balance shaft and, what's more, its left and right banks have been realigned minutely, all the better to suit Buick's split-pin, even-firing crankshaft. The latter, introduced by the Division in 1977, is a clever way to make the V-6 forget it's a 90-degree design like its V-8 siblings.

The Reatta also comes with 4-speed automatic the only choice and, like the Allanté, has generous wheels and tires, suspension several clicks stiffer than the average luxury car, anti-lock brakes and all possible extras included in the base (if such a word applies here) price. The air is automatically conditioned, for instance; the seats go up, down, in and out; there are remote controls for the mirrors, the trunk lid and so forth. The Allanté is billed as America's only ultra-luxury automobile and the only extra listed is a cellular telephone. The Reatta is supposed to be more luxurious than the average sports car and more sporting than the ordinary luxury car, so one can order some extra seat controls and a sunroof, and at some time in the future there will be a convertible version of the Reatta.

If one acts as a critic, taking notes and making professionally artistic judgments, neither car is especially distinguished. There are no strikingly innovative features or touches. We aren't revisiting the tailfins of 1949 or that first Buick Riviera of 1963. If these cars weren't styled by committees, they surely were reviewed by them.

But out in public where people don't know much about art but let you know when they know what they like, the Allanté and Reatta are knockouts. High school kids in souped-up Mustangs hang U-turns in the center of small towns to follow the Reatta and ask informed questions. Men and women wave approval and roll down their windows in traffic to ask what the Allanté is, and how much. This is a sample and not a scientific one. But subjectively, what we saw firsthand says these cars are attractive to the public; i.e., they will do what their makers hope they'll do, namely make a good impression.

The good news continues. All kidding aside, General Motors does know how to build cars, and both of these examples are perfectly good cars.

DOUBLE JACKPOT

WE'D BEEN TESTING the 1988 version of the Cadillac Allanté and Buick's new 2-place Reatta coupe around our Newport Beach, California offices when we began to wonder: Were the approving nods offered to the Allanté or the wide-eyed take-and-doubletake stares gathered by the Reatta indicative of a trend? In Newport Beach, where Ferraris are a dime a dozen and Mercedes outnumber Chevys in registrations, could we rely on such an informal referendum? What was it about this pair of upscale 2-seaters from General Motors?

So we went out in search of some universal truths to that home of the very nouveau riche and not so rich; to where the sizzle counts as much as the steak; to where everyone dreams of hitting the Big One. Could the luxury pair cut it in Las Vegas?

Getting there was half the fun. The plan was to drive halfway, then switch cars. As we nestled into the luxury cruisers for the 275-mile drive, visions of hi-lo splits, $1–$3 razzes and stud and hold-'em-with-joker poker hands danced through our heads. The Reatta and Allanté would be great for casino hopping. Soon we were over the Nevada state line, and not long after that, in downtown Las Vegas.

Somewhere between Ethel M. Chocolate Factory's Famous Free Family Tour, Sassy Sally's Double Jackpots and Ripley's Believe It Or Not Torture Cave, it hit us: Everywhere we went, the cars got noticed—the Caddy for its classic lines, the Buick for its Wow Factor. People tooting, honking, flashing, waving, shouting for us to pull over for a look-see. Definitely a different crowd than the politely waving Newport Beachers.

Talk about unidentified driving ob-

Ask the man who parks one.

jects. "Reatta? What kind of car is that?" asked a young driver of a red, late-model Celica as we cruised downtown Vegas in the bright-red 2-seater at 5:00 a.m. "Wow! A Buick with a TV in it! Looks like my Toyota," he bubbled enthusiastically (it did, sort of). Then spotting the jet-black Allanté bringing up the rear, he exclaimed, "Hey, there's *the* Caddy!" He knew which one.

In fact, a lot of curious onlookers, drawn to a glimpse of Cadillac Style and Buick's latest Great American Road Car, seemed unfazed by the serious coin one would have to put up to own one of these luxury 2-seaters, particularly for the Allanté. Vegas is a *cash* town. And I don't mean just the gamblers. People make a living knowing how much to charge and when to put out their hand. When queried, legions of cabbies, security guards, busboys, bell captains, bartenders, keno runners and slot people knew that the Caddy would set them back a cool 55K.

Even the tourists and second-time-around honeymooners were hip. When asked how much they'd give for that red Buick over there, a middle-aged couple from Idaho looked us straight in the eye and said, "$27,000." Right on the money. Bingo! You win the car. And they were only in town for the rodeo!

But the real savvy operators we wanted to quiz are the guys and gals who get to park all of the cool rides—parking valets for the top hotels and casinos. At the Golden Nugget Hotel and Casino in downtown Las Vegas, we asked valets Greg Swaggerty and Robert Askin to rate (and park) the Allanté and Reatta. We figured these guys get to pilot a lot of luxury cars, if only for a few moments each, but a good cross-section nevertheless. If the Allanté and Reatta were hot stuff, Greg and Robert would recognize it. In all honesty, we half expected both cars to come back with bells, bars and buzzers ablaze, what with the onboard computers and all. But no, our Golden Nugget valets had these cars scoped out.

Greg confessed that it sometimes takes him five minutes to find the lights, seat controls, wipers, radio and so on, but he always gets his man. And both valets liked the size of the two cars; nimble, easy to park, with crisp low-speed response. Yes, even in the parking structure, these guys give the cars their own kind of workout.

On a personal choice, it was a toss up, with Robert giving the Allanté the nod for its classic looks and Greg opting for the more sporty Reatta.

As to the question of which car the parking valets would rather attend to, the answer was easy. The Allanté. Cadillac owners are better tippers.

—*Ron Sessions*

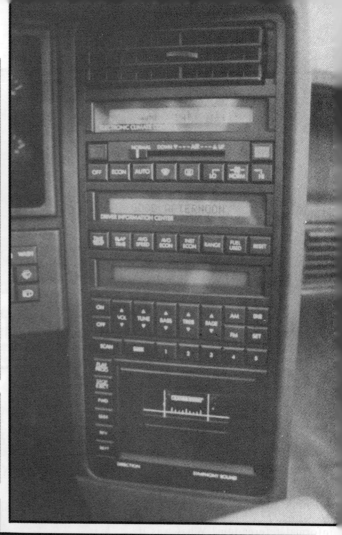

The Allanté design is cool, in the McLuhan sense of hot/cool. There's elegance in its interior, but austerity as well; a collection of taut surfaces and hard edges, sort of a space age Mercedes in spirit. Even the underhood area shows evidence of design, not just engineering.

GENERAL DATA

	Buick Reatta	Cadillac Allanté
Price		
Base price	est $27,000	$56,533
Price as tested[1]	est $27,800	$56,533
General		
Curb weight, lb	3380	3495
Test weight	3530	3680
Weight dist (with driver), f/r, %	65/35	61/39
Wheelbase, in.	98.5	99.4
Track, f/r	60.3/60.3	60.5/60.5
Length	182.8	178.6
Width	73.0	73.4
Height	51.2	52.0
Fuel capacity, U.S. gal.	18.2	22.0
Engine & Drivetrain		
Engine type	ohv V-6	ohv V-8
Bore x stroke, mm	96.5 x 86.4	88.0 x 84.0
Displacement, cc	3791	4087
Compression ratio, :1	8.5	8.5
Bhp @ rpm, SAE net	165 @ 4800	170 @ 4300
Torque @ rpm, lb-ft	210 @ 2000	230 @ 3200
Fuel delivery	elect. port fuel inj	elect. port fuel inj
Transmission	4-speed automatic	4-speed automatic
Gear ratios, :1	2.92/1.57/1.00/0.70	2.92/1.57/1.00/0.70
Final drive ratio, :1	2.97	2.95
Chassis & Body		
Layout	transverse front eng/front drive	transverse front eng/front drive
Body/frame	unit steel	unit steel
Steering type	rack & pinion, pwr-assist	rack & piniofi, pwr-assist
Steering overall ratio, :1	15.6	15.6
Brake system, f/r	10.4-in. vented discs/10.2-in. discs, ABS, vacuum-assist	10.2-in. vented discs/10.0-in. discs, ABS, hydraulic-assist
Wheels	cast alloy, 15 x 6J	forged alloy, 15 x 7JJ
Tires	Goodyear Eagle GT+4, P215/65R-15	Goodyear Eagle VL, 225/60VR-15
Suspension, f/r	MacPherson struts, lower A-arms, coil springs, tube shocks, anti-roll bar/ Chapman struts, lower A-arms, transverse fiberglass leaf spring, tube shocks, anti-roll bar	MacPherson struts, lower A-arms, coil springs, tube shocks, anti-roll bar/ Chapman struts, lower A-arms, transverse fiberglass leaf spring, tube shocks

[1] Price as tested includes, (for the Reatta) std equip. (CRT display with AM/FM stereo cassette, auto. temp control, elect. window lifts, elect. adj mirrors, central locking, cruise control, leather seats & steering wheel, anti-theft alarm), 16-way elect. adj driver's seat ($800); (for the Allanté) std equip. (air cond, ABS brakes, detachable hard top, AM/FM cassette, leather 10-way elect. adj seats, cruise control, leather steering wheel).

The Allanté V-8 is as smooth and responsive as a relatively big V-8 with fuel injection should be. It has been tuned to work where it will live, so the car gains speed and cruises with style, quietness and adequate reserve up to maybe 80 mph; beyond those limits things get strained, but that's not a flaw: Buyers who want to suffer in town so they will be ready to go 200 mph can find such cars elsewhere.

The subtle payoff may be that the Cadillac V-8 has a wonderful sound, an exhaust note that sounds just like the old solid-lifter Chevy V-8 . . . but from a block away. This is a sound from the prospective buyer's childhood, the way electric trains mimic early childhoods.

The Reatta has two fewer cylinders and performs in scale. This too is a responsive, willing engine at normal road speeds, and when there isn't enough torque in high gear, the automatic clicks down one or two in an instant and there the power is, at the expense of some roar. The V-6 has a growl, and it's also entertaining to hear albeit the reliance on lower gears can get tiresome going up a grade.

The Allanté and Reatta don't *feel* at all alike, while they *drive* almost exactly the same. Steering is fast enough and direct enough but lacks feel. The ride is fine, with plenty of wheel travel and control provided by what the engineers believe to be sports-car suspension: It's not, but it is firm enough and supple enough to give both cars just the balance of response and comfort they should have.

The handling comes in one mode: moderate but inflexible understeer. That's it. One goes best and fastest by simply not making any sudden moves. Turn in smoothly, keep the power on and accept the speed dictated by the squeal of the tires, and either car will give a good account of itself.

In dreamland, when one wishes they'd saved the air fare and used the money to build the 4.1-liter V-8 into a 6.2-liter V-12 so the power would have forced them to give a more impressive hood and rear drive, one could come up with a car that would give a good account of itself against, say, a Ferrari 412. Instead, what one has here is a car that will be lunch meat for a Porsche 928. And the Reatta will be left in the lurch (oh, clever!) by a passle of cars.

Back in real life, the Allanté has some structural problems. One is almost frivolous: The procedure for removal of the hard top is awkward and overly elaborate, the same goes for the soft top; and because the car is a convertible, it comes with cowl shake and door shake. The Allanté is heavy for its size, supposedly to make it stiff and all-of-one-piece, but it's not. The driver can feel the body working against itself on bad roads or even across speed bumps if they're hit at an angle; heck, you can even *hear* the body creak under stress.

The enclosed Reatta is better, as it should be, although it's difficult not to expect the same shortcoming when the Reatta convertible arrives in a year or so.

Both cars have anti-lock brakes and, of course, come with power assist, and both sets of brakes are oddly rewarding—not merely in performance, as both stop in respectably short distances, without fade and with anti-lock at the ready if things get non-sticky. The surprise is that neither system has feel when the pedal is first depressed; but then, if you push down harder, the brakes come into play geometrically rather than in direct ratio, so the feeling is, *My goodness but these brakes really work! They are on my side!* And so they are, but it does take some getting used to.

So do the interiors, each in its own way.

The Allanté is cool in the McLuhan hot/cool media theme. There are smooth, taut surfaces, lots of hard edges and the feeling is one of elegant austerity. The steering wheel is too big, evidence that the target was spelled M-B.

The seats adjust in every direction there is, electrically, and one can punch the memory button and have all the settings

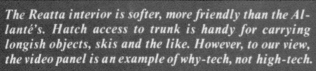

The Reatta interior is softer, more friendly than the Allanté's. Hatch access to trunk is handy for carrying longish objects, skis and the like. However, to our view, the video panel is an example of why-tech, not high-tech.

> *Even the rich need to know they could succumb to temptation if they chose to.*

PERFORMANCE

	Buick Reatta	Cadillac Allanté
Acceleration		
Time to distance, sec:		
0–100 ft	3.3	3.3
0–500 ft	8.9	9.2
0–1320 ft (¼ mi)	16.8	17.1
Speed at end of ¼ mi, mph	80.7	80.0
Time to speed, sec		
0–30 mph	2.7	3.0
0–60 mph	8.9	9.3
0–80 mph	16.5	17.1
Estimated top speed, mph	125	119
Fuel economy, mpg	16.6	22.4
Braking		
Stopping distance, ft, from:		
60 mph	155	168
80 mph	273	275
Pedal effort for 0.5g stop, lb	14	na
Brake fade (six 0.5g stops from 60 mph), % increase in lb	14	na
Control	excellent	na
Overall brake rating	very good	very good
Handling		
Lateral acceleration, g	0.80	0.77
Slalom speed, mph	62.1	61.0
Interior noise, dBA¹		
Idle in neutral	49	56/54
Maximum, 1st gear	73	69/70
Constant 30 mph	59	62/56
50 mph	64	66/64
70 mph	72	71/68

¹For Allanté: interior noise with soft top/hard top.

CALCULATED DATA

	Buick Reatta	Cadillac Allanté
Lb/bhp (test weight)	21.4	21.6
Bhp/liter	43.5	41.6
Engine rpm @ 60 mph in top gear	1650	1600
Brake swept area, sq in./ton	175	196

return to where they were before the parking valet, the one who left the radio on the drum machine/punk rock and reggae station, messed it up.

The Reatta is a softer, more friendly car on the inside. Again a 2-seater, again plenty of room for those two and all the luggage one can imagine lugging up the stairs at that quaint little inn. The wheel is smaller and the doors smaller and lighter or perhaps the door and seat relationship is better, because the Reatta is more easily clambered in and out of. The Reatta's leather-covered seats come with basic adjustment (in fact, the quickest acting in our experience) and with an optional second set of controls. Get them. They let you set the seat cushion to just where you want it and then vary the seat back, settings the regular controls combine. The Reatta has less mechanism aft of the seats, no folding top yet, and thus better rear vision.

Our Allanté's basic instruments are analog, round dials right there in front, just as they should be. This is new for 1988, a no-cost option in lieu of the digital/analog hybrids of the original design.

The secondary systems, as in radio and air and the obligatory computer to tell you what time it is and the outside temperature and how many gallons of fuel you've used, are digital, housed in a panel in the center of the dash. They have plenty of room and the buttons are of generous size, what used to be called man-sized.

This panel plays the role of silent butler. "Good Morning" it spells out when you turn on the key; then it checks all the instruments, etc, and tells you it's ready to go. The greeting becomes "Good Afternoon" when appropriate, and as dusk approaches the panel gently suggests that you may wish to turn on the headlights, sir. All in good taste and good fun.

The Reatta is much too much in the other direction. Buick developed a fully electronic, full-digital, touch-sensitive multiphase CRT panel for its Riviera, and because it cost so much, Buick engineers are under orders to use it wherever they can. Such as in the Reatta.

First, it's all digital and harder to read at a glance.

Second, and more critical, the driver is no longer allowed to use the sense of touch. There are no knobs or buttons, so you can't simply adjust temperature or volume by feel, with eyes on the road. You have to look at the panel to use it.

Third and last, this blithering surfeit of fact goes against the car's intended character. The Reatta is supposed to be luxurious, the data panel is high-tech and Buick can't tell one from the other.

It's easy. High-tech is when you *cook* great meals because you have a microwave, while luxury is when you *eat* great meals because you have a chef....

And Grand Touring means you don't play video games.

So, to summarize: Judged on the basis of being cars—things to let you go places and do things, in comfort—the new Reatta and its corporate sibling Allanté are very good. They ride nicely, have more than enough power for any legal purpose, steer and stop predictably and so forth.

The worry here is these aren't supposed to be just cars. They are supposed to go beyond that, just as their prices go beyond what anybody needs to pay for a way to drive to school or shop or office.

These are supposed to be superior machines and superior statements, and they are merely nice cars.

We're thankful that their design teams avoided overdoing things the wrong way, as in any of those horrible things festooned with gold plate and fake landau bars.

But in doing that, they missed overdoing things the right way, the way Mercedes and Ferrari overdo things by giving the buyers more power, more speed, more *something* than they'll use. Even the rich need to know they could succumb to temptation if they chose to.

For some reason, it's become fashionable to take potshots at Cadillac. After decades of being almost synonymous with the word "class," Cadillac in the early '80s had come to be just a shadow of its former self. Awash in the uncertain seas of the General Motors reorganization, Cadillac division offered its status-seeking customers cars that were simply lightly rehashed versions of other, lesser divisions' models. And on those infrequent occasions when Cadillac ventured out on its own (with the infamous 4-6-8 engine, for example), it seemed to run into nothing but disaster. On one side, the division saw its market share being eroded by foreign competition that had staked out the high ground; and on the other, it saw market share being gobbled up by domestic competitors who retained the traditional American front engine/rear drive configuration long after Caddy had set about downsizing and shifting to front-wheel drive. Frankly it seemed like Cadillac just couldn't win.

But now, at long last, Cadillac has a winner. It's a car that harks back to the glory days when the most exclusive Caddys were fitted with custom bodywork from the best coachbuilders in the world, and at the same time speaks of a new youth and vibrancy long missing from the Cadillac model range. It's a car that makes a statement of style, quality, and enjoyment of the driving experience. Its name: Cadillac Allanté.

Sure, you cynics might be saying, isn't the Allanté simply a convertible version of the none-too-daring Eldorado and Seville model range that's hardly setting the sales charts on fire? Isn't it a close cousin to the Olds Toronado and Buick Riviera? Isn't it just a Buick Reatta clothed in Italian steel with a $20,000 higher pricetag? The fact is, if we were to say yes to all these questions, we'd only be telling a half truth because, while the Allanté does share some underpinnings with its lesser siblings, its essence and feel are quite different from the run-of-the-mill GM offerings of late.

What we really see in the Allanté is a sincere attempt to bring Cadillac back to the well-deserved high status the division once enjoyed. "Sincere" is the operative word here. The Allanté is not, like so many recent GM products, a low-tech engine stuffed into old-before-its-time bodywork with all the ergonomic convenience of a left-handed monkey wrench. It

MOTOR TREND ROAD TEST
Cadillac Allanté
The comeback kid

PHOTOGRAPHY BY MIKE BANKS by Jack R. Nerad

would've been easy for Cadillac to take unsold Eldorados (it has plenty), peel off the tops, stick some designer labels on the side (Giorgio Linguine? Perry Mason? Willie Wonka?), and tell the public a major breakthrough has been achieved in luxury motoring. But, to its credit, Cadillac didn't do that.

What Cadillac *did* was create an honest car with honest features that are truly better than those found on less-expensive models. Just as an example, look at the Allanté's engine. Think how convenient (and how cheap) it would have been for the Cadillac folks to insert their current V-8 (even their much-improved 4.5-liter), utter yet again the time-worn phrase "luxury buyers don't care about performance," and let it go at that. But they didn't. Instead they went to the considerable work and expense of creating a sequential port fuel-injection system and tuned intake manifold for their 4.1-liter V-8 to up its output from a meager 130 hp to a solid 170 hp. Torque is way up as well to a substantial and satisfying 235 lb-ft at an easily attainable 3200 rpm. From a standing start, 60 mph

The Allanté speaks of vibrancy long missing from the Cadillac model range

arrives in less than 10 sec. Sure, this isn't a motor that'll light Don Garlits' fire, but it definitely won't leave you dawdling behind everyone else at every stoplight.

Backing up the transverse-mounted aluminum V-8 is a considerably revamped version of GM's lockup 4-speed automatic, dubbed the THM F7. The transaxle, beefed up to deal with the engine's higher torque output, controls shifts electronically rather than simply through hydraulics, and utilizes a viscous converter

clutch. With its electronic controls, the F7 gives crisp, quick shifts without being jarring. There's no uncomfortable lag in kickdown shifts, either, though with the torquier motor, fewer downshifts are required than in '87 model Eldos and Sevilles.

Our only problem with the trans, a minor one, came on a few cold-start 1-2 shifts when we experienced a slight *ker-thunk*. After much head-scratching, the culprit was discovered to be an intermittently sticking valve. This discovery came after Cadillac removed the transmission and shipped it back to Detroit for diagnosis. Given the mild nature of the complaint, we thought the whole procedure might have been a bit of overkill done for us because we're in the car magazine business, but Cadillac assured us *all* THM F7s that experience problems are shipped to Detroit for analysis, since the trans is so new and incorporates technologies that will likely be used in future GM transmissions. We were also assured the problem ratio on the new trans is low.

Our guess is that few drivers will have a problem with the Allanté's handling, either. This is a car that

Although designed as a convertible, the Allanté comes with a standard bolt-on aluminum hardtop that transforms it into a handsome coupe.

takes to the road like a fine imported touring sedan. That's quite a compliment considering the fact that the Allanté's sleek aluminum roof comes off when you'd like to shift into convertible mode. (Of course, a power-operated soft top resides behind the back seat when you'd like to drive top up and have forgotten your hardroof at home.) As you're no doubt aware, many convertibles on the road these days are more shakers than movers, their structural integrity emasculated by the removal of their precious tops. Not so the Allanté. There's no evidence of cowl shake and no evidence of each wheel wanting to take its separate path to the same destination. Instead, the Allanté's handling is precise, and, despite its 3500-lb curb weight, the car feels agile. Imagine a Cadillac that feels at home on curvy mountain roads. (Hey, it might take some doing, but try.)

The reason the Allanté feels so at home in the twisties is the homework done by the Cadillac suspension engineers. The suspension is all independent—Macstruts up front and a composite transverse leaf spring at the rear. All this is just like the Eldo/Seville. The magic in the formula is performed with spring rates, shock valving, and an excellent set of tires. The Allanté rides on P225/60VR15 Goodyear Eagle VL tires designed specifically for the car. All this tuning contributed to a lateral acceleration figure of 0.79 g, good transient behavior, and confident roadholding. Adding to the confident feel are the vented front and solid rear disc brakes equipped with the extremely sophisticated Bosch ABS III.

The Allanté is a sincere attempt to bring Cadillac back to the well-deserved high status it once enjoyed

Equally confidence-inspiring is the car's interior. In fact, we've rarely experienced an interior that *works* as well as the Allanté's. It's unfortunate that when the car was introduced, it could only be had with a garish vacuum fluorescent dash layout, because the optional-at-no-extra-cost analog display available for '88 is topnotch. Large round dials display vehicle speed and engine rpm, while easy-to-read but smaller secondary dials help you keep track of the charging system, water temperature, and oil pressure. A self-diagnostic system dubbed the Driver Information Center fills you in on possible failures plus calculated fuel economy, range, average speed, and other tidbits of trivia. The placement and operation of the pushbuttons and controls are universally excellent and intuitive. No wading into the manual is needed to turn on the interior lights, adjust the heat, or activate the windshield wipers.

The Allanté's seats don't just simulate Recaros, they are Recaros, and they power adjust 10 different ways. (Count 'em.) The seat controls, located on the outside edge of the seat out of sight, are a bit difficult to get used

The Allanté's stylish interior features leather-wrapped Recaro power seats and can be fitted with a first-rate analog instrument package.

to, but once dialed in, the seats are truly superior, clothed in hand-matched Cogolo leather. The convenient pass-through between the trunk and stowage area behind the seats takes no getting used to at all.

The car's exterior styling isn't hard to like, either, but that's both good and bad. First off, let's make it clear we think the Allanté is an attractive shape that looks even better in "real life" (or at least our California approximation of "real life") than in photographs. We'd venture to say, however, that it's a bit of a shame the shape (drawn and built, as you know, by Pininfarina) isn't more distinctive. We figure the guy who ponies up in

It's reminiscent of when Caddys were fitted with custom bodywork from the world's best coachbuilders

excess of 50 big ones for a car wants to be noticed, and, while the Allanté does turn some heads (particularly in red), it doesn't stop traffic.

There's no doubt, though, the exterior design works well. The drag coefficient is a commendable 0.34, but, even more important, in top-down driving, wind buffeting in the passenger compartment is kept to a minimum. (It'd be a shame to mess up those $50 razor cuts with some unruly wind.)

All in all, despite some niggling little fit-and-finish problems (undoubtedly familiar to owners of Italian cars) and a price that might give prospective buyers pause (if not palsy), we think the Allanté is a winner. Certainly, it has gotten off to a slow start in sales, but remember the Corvette didn't exactly set the sales charts ablaze in its first year, either. We think the '88 Allanté is far better than last year's model, and we're convinced that it puts Cadillac on the road back to prominence. **MT**

TECH DATA

Cadillac Allanté

GENERAL
- Vehicle mfr.: Cadillac Motor div., General Motors Corp., Detroit, Mich.
- Body type: 2-pass., 2-dr.
- Drive system: Front engine, front drive
- Base price: $56,533
- Price as tested: $56,533

ENGINE
- Type: Transverse 90° V-8, water cooled, aluminum alloy block and heads
- Displacement: 4087 cc (250 cu in.)
- Compression ratio: 8.5:1
- Induction system: Electronic fuel injection
- Valvetrain: OHV, 2 valves/cylinder
- Max. power (SAE net): 170 hp @ 4400 rpm
- Max. torque (SAE net): 235 lb-ft @ 3200 rpm
- Emissions control: Closed-loop dual-bed monolith
- Recommended fuel: Unleaded premium

DRIVETRAIN
- Transmission: 4-sp. auto.
- Transmission ratios
  - (1st): 2.92:1
  - (2nd): 1.57:1
  - (3rd): 1.00:1
  - (4th): 0.70:1
- Axle ratio: 2.95:1
- Final drive ratio: 2.07:1

CAPACITIES
- Crankcase: 7.3 L (7.68 qt)
- Fuel tank: 83.2 L (22 gal)
- Luggage: 368 L (16 cu ft)
- Range (at EPA combined): 563 km (352 mi)

SUSPENSION
- Front: Independent, MacPherson struts, coil springs, anti-roll bar
- Rear: Independent, MacPherson struts and transverse composite leaf spring

STEERING
- Type: Rack and pinion
- Ratio: 15.6:1
- Turns (lock to lock): 3.2
- Turning circle: 11 m (36.1 ft)

BRAKES
- Front: 260 mm (10.3 in.), vented discs
- Rear: 254 mm (10 in.), discs
- Anti-lock: Bosch ABS III

WHEELS AND TIRES
- Wheel size: 15 x 7.0 in.
- Wheel type: Forged aluminum
- Tire size & construction: P225/60VR15VL
- Tire mfr. & model: Goodyear Eagle VL

DIMENSIONS
- Published curb weight: 1587 kg (3496 lb)
- Weight distribution, f/r: 61/39%
- Wheelbase: 2525 mm (99.4 in.)
- Overall length: 4537 mm (178.6 in.)
- Overall width: 1866 mm (78.5 in.)
- Overall height: 1327 mm (52.2 in.)
- Track, f/r: 1533/1533 mm (60.4/60.4 in.)

SPECIFICATIONS
- Power-to-weight ratio: 20.5 lb/hp
- Drag coefficient: 0.34
- EPA (combined): 20 mpg

MEASURED PERFORMANCE
- QUARTER MILE (TIME): 17.21 sec
- QUARTER MILE (SPEED): 80.8 mph
- BRAKING
  - 60-0: 150 ft
  - 30-0: 46 ft
- SKIDPAD: 0.79 g

SPEEDOMETER CALIBRATION

Indicated	30	40	50	60
Actual	29	40	50	60

ACCELERATION (SEC)
- 0-30: 3.34
- 0-40: 4.90
- 0-50: 6.95
- 0-60: 9.47
- 0-70: 12.61

FIVE EXOTIC CONVERTIBLES

Whining through the gears in California's Napa and Sonoma

TOPS THAT COME off can forgive a multitude of automotive sins. Take a temperate clime, a sunny day and a twisty road through interesting country, and even a mundane car can seem altogether tolerable provided your view overhead is unrestricted. So imagine our enthusiasm for this particular comparison test: five exotic convertibles—the Cadillac Allanté, Chevrolet Corvette, Ferrari 328 GTS, Mercedes-Benz 560SL and Porsche 911 Cabriolet—on a journey to, from and around California's Napa and Sonoma wine country.

Not that these cars haven't been scrutinized within fairly recent R&T memory. The Allanté was fresh from its March 1988 table stakes game with the Buick Reatta in Las Vegas. A soft top Vette was one of our 40th-anniversary test subjects in June 1987. The Ferrari 328 GTS didn't differ remarkably from the one we evaluated in May 1986 when it was introduced. The Mercedes SL got its boost in displacement along with the 560 designation at about the time of our March 1986 road test. And it was January 1986 when we tested everything Zuffenhausen built, including the 911 Cabriolet.

So why go gallivanting off north with these cars now?

Uh, that wasn't a gift subscription, was it?

We assembled early one morning just north of the Los Angeles sprawl, at Interstate 5's Grapevine. This long climb to 4144 ft at the Tejon Pass can still be formidable for automobilists, but this time around none of our high-powered quintet even breathed especially heavy. True, the Allanté had its share of automatic shifting fits between 4th overdrive and 3rd direct, but this was easily eliminated by dropping down into 3rd manually. The other automatic of our group, the Mercedes, burbled along on sheer 5.5-liter torque. The quasi-sometimes-automatic, our Vette 4-speed+3, could be tickled back and forth from overdrive to direct with gentle nudges of the toe.

PHOTOS BY RON PERRY

COMPARISON R&T ROAD TEST

Not exactly your ordinary parking lot: Our five exotics repose awhile in the shade of Chateau Montelena.

The Ferrari, relatively short geared running 3100 rpm at 60, could handle just about any Interstate in 5th alone. Though also possessing good torque, the 911 preferred 4th for whiffling past slower traffic.

Down into the San Joaquin Valley, we headed up California Route 33 through Taft and Devil's Den toward Coalinga, then onto Routes 198 and 25 through Hollister. This itinerary starts off as an arrow-straight 2-lane amid stark landscapes dotted with oil derricks, but eventually gives way to decidedly more scenic country. Top-down aromas improve as well, from penetratingly raw hydrocarbons to rich agriculture, epitomized by olfactory celebration of Gilroy, the Garlic Capital of California. And, one thinks, perhaps of the world.

Other cars along the way were few, far between and most cooperative, giving us good opportunity to evaluate such automotive virtues as directional stability, wind protection and stereo fidelity at full volume. Going really quickly, the 328 GTS displayed its *pur sang* in feeling rock solid right up to its top speed. The Vette had the power to pace the Ferrari and its seating position protected the driver from the wind blast, but more than one of us thought things got a bit darty at really elevated velocity. Cockpit buffeting and what felt like an aerodynamic wall defined a comfortable upper limit for the SL's cruising. By contrast, drivers felt cosseted by the Porsche's low seating position. Last and slowest, though by no means embarrassing itself in this twice-Double-Nickel-plus company, the Allanté flat ran out of horsepower about the time cockpit buffeting started getting out of hand. And any one of them, used imprudently, could have earned us one helluva citation.

California Route 25 skirts the Diablo Range and it's made up of smooth sweepers of various speeds. We thought we knew these cars fairly well going in, and indeed a couple of truisms were confirmed along this stretch: First, there's the dichotomy of sports cars—Corvette, Ferrari and Porsche—versus GTs—Allanté and Mercedes. It's no slight to either group, just a recognition of avowed purpose. Second, within the sports category, a clear distinction separated the Vette from the other two.

Steering, brakes and suspension of the Ferrari and Porsche were designed with this road in mind. Control communication and response to driver input were predictably loud and clear. At one's own comfortable 8/10ths, these cars were absolute delights. Probe harder, though, and one had better be ready to contend with the inevitable trailing-throttle oversteer accompanying rear weight bias.

By contrast, the Corvette's behavior at 8/10ths was less fun. Its controls didn't telegraph grip as well; the steering felt overly sensitive; the brakes were less amenable to easy modulation. In a corner, the car responded with an occasionally disconcerting two-step: turn-in/then plant. But—and this is an immense qualification in the Vette's favor—it was the most forgiving car of the group, sports and GTs altogether.

Whereas the Ferrari and Porsche felt as though they might snap back on the careless, the Corvette's quirks got no more pronounced at its limit. As one driver said, "The Corvette is idiot-proof, even when out of control."

Our two GTs acquitted themselves handily through the twisties as well. The Mercedes proved surprisingly nimble, with its V-8 torque helping to balance a chassis that exhibited just a tad of float now and again. The steering gave the impression of being on the slow side, but factor in as well Mercedes' decidedly oversize wheel.

Cornered throttle-on and moderately hard, the Allanté displayed the push one would expect from front-drive weight bias. But, at speed, a quick liftoff of throttle produced a rapid tuck-in, probably caused by the relatively soft suspension transferring even more weight to those hard-working front tires. Used judiciously, this characteristic could tighten a line very nicely.

North of Gilroy, our route took us onto 101 through San Jose and then into San Francisco. Careful planning brought us there right at the evening rush hour.

We do these things for you, Gentle Reader.

What did we learn inching along toward the Golden Gate Bridge? The Ferrari's shift linkage was heavy and notchy (it was only later we confirmed its really serious flaw); the Porsche's shifter was much improved over the wobble-sticks of 911 worst-case memory; the Vette's shifter felt as if it were actuated by particularly loose-linked chains. And the other two guys really lucked out drawing the automatics for this leg of our journey, though the Mercedes driver complained of an overly stiff accelerator spring. Isn't there always something?

We reached our destination, the really lovely Vintner's Inn outside Santa Rosa, reasonably bright-eyed and bushy-tailed, though in truth no one suggested taking a little drive before dinner. Buttoning up each of our quintet for the night gave us added opportunity to assess the various ways of making an open car.

The Ferrari is the most straightforward, though really only a targa, not a classic convertible. When not latched in place, its rigid top stows vertically behind the seats.

The Porsche's soft top is the most automatic, with optional ($2524!) electric actuation even to the front bow's screwing itself down onto the windshield header. Stowed, however, the top sits high enough to restrict rear vision for some.

The Corvette's top works best of the manual variety. A cockpit switch (actually, any one of three) releases the rigid cover. The latches are large and easy to lock. When stowed, the top resides low enough not to alter the swoopy lines of the car's rear deck. However, it also takes up most of what passes for the Vette's luggage space.

The Mercedes' top is the most classic, being purely mechanical. Its rigid cover is released via a cockpit-mounted crank, the top is pulled up manually, its front bow is locked in place with two twists of a separate tool, and its rear portion is latched down with another motion of the crank. Though Mercedes hardware is of high quality, here the design is simply behind the times. Of the four soft tops, for instance, the SL's beats and billows the most when erect. Left at home, by the way, is a sturdy, if heavy, hardtop for prolonged lidded motoring.

The Allanté also has a hardtop to leave at home. And, by contrast with the Mercedes, its soft top buttons things up especially well. However, the Allanté's retraction mechanism is one of those artful designs that's probably impossible to do correctly in mass production. Everything depends on perfect alignment and high-quality components, neither particularly evident in our almost-new test car. One time, we managed to snag the retracting rear-window portion on its own electrically actuated latch; it took a half hour of fiddling to bring things right. Other times, the roller-in-slide arrangement locating the top's rigid cover slipped its grooves, giving us visions of its popping up at the most inopportune time. Though it rattled a lot, the Allanté's cover never actually attempted this 300SLR imitation.

One of our trippers preferred driving all the cars with their tops up. Another wanted them up on freeways but down through the twisty stuff. Two others insisted on tops down, whatever the conditions. And one wimp based his choices on sun angle and ambient temperature. What with our usual swapping cars and all, it wasn't as though we didn't get plenty of practice with these convertibles.

The whole point of the next day's driving was to enjoy the scenery and roads of Napa and Sonoma, not to say the fruits of a vineyard here and there. An adjacent sidebar by our staff oenophile (Greek: *oinos*, wine + *philos*, loving) focuses on this last bit. Suffice it here to stress that our day was a sensible affair of good driving, food and wine, roughly in that order.

California Route 12 is the Sonoma Valley's principal artery,

Route 29 is Napa's, a healthy range of good-size hills separates them, and here and there roads snake over the hills. Our favorite begins as Trinity Road, to the east off Route 12 just south of Kenwood (the latter, the home of Kenwood Vineyards, after which a staff member's Husky/Malamute is named). Trinity Road leads onto Dry Creek Road, then to Oakville Grade as it gets positively serious about the descent. It's all twisty 2nd and 3rd gear, with more than a few of what the French call *lancettes* and we call switchbacks. A loop off this route passes by vineyards so select that they welcome tours (by appointment only, please) but their limited production precludes wine tasting in the usual sense.

It was along Trinity Road that the driver who drew the Ferrari confirmed this car's most serious shortcoming: a linkage whose full-throttle 1st-2nd shifts led occasionally to a blind gate. One of the other drivers recalled similar problems with other Ferraris, caused, he conjectured, by the engine being torqued over on its mounts and misaligning the linkage. Whatever the problem, we all agreed it was inexcusable in a car of this caliber.

North of Calistoga, Route 29 passes Chateau Montelena, a particularly beautiful winery whose main building—a chateau nestled in the hillside—was built of stone quarried in Europe. Continue up Route 29 and there's a great climb with sweeping curves and magnificent views of the entire Napa Valley from a couple of well-placed turnouts. We found it particularly beautiful at dusk.

Back at our lodgings, we enjoyed a bottle of the Inn's own vintage and pondered the significance of it all. There's not a bad car in the lot, we concluded. In fact, one of the drivers had his wife fly up to spend the weekend, and we all had the devil of a time helping him decide which car to retain. We finally settled on the Ferrari, despite his having to empty his clothes loose into its carpeted rear stowage because the suitcase wouldn't fit.

The Ferrari wasn't a universal choice, though. And as the chatter progressed, we identified two distinct scenarios: a car for touring California's wine country (or anyone else's, for that matter) and a car for daily use back at home. Nearby are isolat-

GRAPE EXPECTATIONS

FINE CARS, FINE wines. The two are synonymous. Both are made to be appreciated and enjoyed. With wine, nothing heightens this experience more than sampling the nectar at its source. Thanks to the wonders of nature, California and other states have regions where the climate is right for the growth of *Vitis vinifera*, the traditional wine grape that is the source of all classic wines. The Golden State has many such areas, but two of the best are the Napa and Sonoma valleys. Some of the best wine in the world is produced here (in the oft-discussed Paris tasting of 1976, a California Chardonnay, Chateau Montelena, and a Cabernet Sauvignon, Stag's Leap, soundly trounced the legendary whites and reds of Burgundy and Bordeaux). What's even better, it's presented to visitors at the 100 or so wineries that dot this piece of California landscape just north of San Francisco.

Most wineries offer tasting, about which more in a moment. But many of them also give tours. True, you don't have to see or know how wine is produced to enjoy it, but your appreciation of this beverage will be enhanced once you've seen how exacting the process is. To find out who does what (and where), you might consult a pocket tour guide such as *California Is Wine Country*, available free from Wine Institute, 165 Post St, San Francisco, Calif. 94108; 415 986-0878. For members of the AAA, *California Winery Tours*, a 62-page booklet produced by the Automobile Club of Southern California, 2601 S. Figueroa St, Los Angeles, Calif. 90007, gives the location, hours and other specifics for many of the state's 726 bonded wineries. There are also several books on California wineries, and some of these publications list the kinds of wines produced by each vintner.

If you've never seen wine in the making, there are hundreds of personable folks who will be delighted to demonstrate how it's done. Almost any time of the year is a good time to visit wine country because the work goes on year round. However, October or November are best if you want to see the grapes picked, crushed and poured into fermenting tanks. A word to the wise: This is peak season in wine country (you think you're the only one with the same idea?) and thousands of wine buffs from what seems like all of northern California will be in Napa and Sonoma, trying to get from Chateau Hither to Yon Cellars, simultaneously, on mostly 2-lane country roads. So, go on a weekday.

Unless you collect winery tours like some of us collect, say, model cars, a visit to one large and one small producer should provide enough background on the subject. Sure, on a hot day in Saint Helena, it's refreshing to duck inside a winery where it's always cool. But unless you have a fondness for the smell of oak barrels and acetic acid, there's little to be learned by trudging through every facility in the valley. If you don't know where to begin, here are my criteria for choosing a winery to tour:

Staff oenophile, left, promotes appreciation of Vintner's Inn Sauvignon Blanc.

ed our final tallies, but like any good philosophical discussion, this one deserves some amplification.

Perhaps another bottle of the Sauvignon Blanc, please.

The Porsche 911 fared remarkably well, even with the professed non-Zuffenhausenfreaken among us. "Exotic without being a nuisance about itself," said one driver. "It's reliable, fast, agile and fun," said another. "The best all-round sports car," opted a third. The 911 was a runaway first for daily use and just a pip behind the Ferrari for exploring amid the vines.

The ultimate accolade for an exotic, the Ferrari topped the wine country tour list, yet finished last around town. "Over the quiet hiss of growing grapes," mused one driver, "you can hear the Ferrari's engine, its gears and people talking about you." Or, to cite the view of its weekend driver, "It's the most visceral and entertaining of the bunch. What's more, it looks as if it belongs in a warm, sunny place full of vineyards."

The Ferrari was not without dissenters, however. "Maybe Walter Mitty would enjoy the Fazzaz in the wine country," conceded one of them, "but a shy little Polish kid would be embarrassed." And taking her back home was even more problematical: "I'd be scared to leave it unattended," noted one of our drivers.

Finishing surprisingly strong, second around town and tying for third amid the vines, was the Mercedes. Surprisingly, we say, because in general demeanor the SL is the most subtle of the bunch, and our particular group isn't always known for subtlety. "But it exemplifies class," said one driver, "and it's the least ostentatious of an ostentatious fivesome." The driver who fought San Francisco's rush hour in the car recalled other virtues: "It has torque aplenty to scoot away from lights, and the automatic is great for traffic jams." Another summed it up for all of us when he noted, "and it's a vault."

The Allanté tied for third in the wine country, finished fourth in daily use and won a few unsuspecting hearts in the process. "Easy to live with," said one driver, "and exclusive at this point because its virtues have yet to be discovered by the general public." Also in its favor were "fresh styling, a heavenly body of the Italian persuasion," as one driver waxed elo-

---

Esthetics: Does it look like a winery, a French chateau, let's say? Or does it resemble a hazardous waste disposal facility? Many new wineries place more emphasis on product than image, and while they produce some magnificent wines, they are quite sterile. And there's nothing that makes the process more uninteresting than staring at rows of stainless-steel tanks in a corrugated steel and concrete building.

Reputation: Is this a winery that has earned the respect of oenophiles? Or is it a wine factory producing nothing more than plonk? Put another way, if it were building cars, would they be Ferraris or Fiats? And which ones would you prefer to see being made?

Product quality: After you have watched wine being made, you'll probably want to taste the stuff, so choose a vintner whose product you enjoy or respect (when in doubt, aim high). Yes, I know a professional taster spits the wine out, but most of us swallow it. So why not enjoy the best?

Ambiance: Because this area is very scenic, many wineries are perfect places to have a picnic (with wine, of course), enjoy the view, or just tarry and prolong your enjoyment of a day in the wine country.

If you still haven't a clue as to which winery to visit, here are some more suggestions. If you don't know a lot about wine (or care less), start with a big winery. Beringer is a good bet because of its limestone caves carved into the hillsides, its ornate tasting room in a 100-plus-year-old Victorian home (the Rhine House) and its good wines. Sterling is another possibility because of its unique setting (to get to the winery, you ride an aerial tram) and its state-of-the-art facilities. Inglenook, Christian Brothers, Robert Mondavi and Beaulieu, located in the same vicinity, are colorful and make good wines, although they are not as visually distinctive (no caves, no trams). Also, Buena Vista and Sebastiani in Sonoma are interesting because they are two of the oldest (hence, quaint-looking and historically significant) wineries in California.

In the same, large-scale category are two champagne producers, Domaine Chandon and Gloria Ferrer. Both facilities look the part, are set up for tours and make an excellent product (Domaine Chandon also has an excellent restaurant on the premises).

Offering a more personalized approach to touring are several small (okay, smaller than Gallo) wineries with big reputations. Chateau Montelena is lovely to look at (be sure to view the swan lake) and continues to live up to the reputation it earned 15 years ago in Paris. Chateau Saint Jean (pronounced Jean, not John), located in the Sonoma Valley, also looks and tastes great.

Then there are the boutiques, the verr-y small wineries producing verr-y special wines in limited quantities. Most of these are not set up for tours, but some have them—by appointment only. It helps to know something about wine if only to gain admittance and to avoid making a total fool of yourself in some guy's kitchen. Zeroing in on a boutique winery is really subjective. (Translation: Your choice is as good as mine.)

Although, after one tour, you may never want to set foot in another winery again, you certainly won't want to pass up tasting. In fact, this is the reason most of us visit wine country—to sample products one might not be able to find elsewhere. Anyway, practically every winery offers tasting which takes no experience at all. Because most visitors are neophytes, most wineries have a program that lets you sample a cross-section of their wines—say, a white, red and blush, unless they make only whites or reds, in which case they'll run through a range of these. Don't expect to turn tasting into Happy Hour. In California where the onus is on the host (don't let friends drive drunk), wineries serve only an ounce or so of wine and let you sample no more than three or four kinds at a time. Some impose a nominal charge for tasting not only to discourage excessive drinking, but also to weed out derelicts. My experience has been that the more you know about wine, the more accommodating a winery host will be.

Given a choice, I prefer to sample only whites or only reds at a single tasting. If conditions allow (a slow day, with me the only person in the tasting room), I may ask to try only a specific variety, say Sauvignon Blanc or Chardonnay, from either different vineyards or different vintages. Of course, I feel compelled to buy a bottle or three, if only to show my appreciation for the winery's hospitality and my enthusiasm for its wares.

And there's certainly a lot to be enthusiastic about in Napa and Sonoma. There's great wine to go with great scenery and great roads.—*Joe Rusz*

GENERAL DATA

	Cadillac Allanté	Chevrolet Corvette Convertible	Ferrari 328 GTS	Mercedes-Benz 560SL	Porsche 911 Cabriolet
Price					
Base price	$56,533	$34,820	$78,850	$62,675	$52,895
Price as tested[1]	$56,632	$38,352	$78,850	$63,975	$60,576
General					
Curb weight, lb	3495	3340	3170	3570	2940
Test weight	3645	3490	3340	3720	3090
Weight dist (with driver), f/r, %	61/39	50/50	44/56	52/48	42/58
Wheelbase, in.	99.4	96.2	92.5	96.7	89.5
Track, f/r	60.5/60.5	59.6/60.4	58.0/57.8	57.7/57.7	53.9/34.3
Length	178.6	176.5	168.7	180.3	168.9
Width	73.4	71.0	68.1	70.5	65.0
Height	52.0	46.4	40.1	51.1	51.6
Fuel capacity, U.S. gal.	22.0	20.0	18.5	22.5	22.5
Engine & Drivetrain					
Engine type	ohv V-8	ohv V-8	dohc V-8	sohc V-8	sohc flat-6
Bore x stroke, mm	88.0 x 84.0	101.6 x 88.4	83.0 x 73.6	96.5 x 94.8	95.0 x 74.4
Displacement, cc	4087	5733	3185	5547	3164
Compression ratio, :1	8.5	9.5	9.2	9.0	9.5
Bhp @ rpm, SAE net	170 @ 4300	245 @ 4300	260 @ 7000	227 @ 4750	214 @ 5900
Torque @ rpm, lb-ft	230 @ 3200	340 @ 3200	213 @ 5500	279 @ 3250	195 @ 4800
Fuel delivery	elect. port-inj	elect. port-inj	elect. port-inj	mech port-inj	elect. port-inj
Transmission	4-sp automatic	4-sp manual + (OD)	5-sp manual	4-sp automatic	5-sp manual
Gear ratio, :1, 1st	2.92	2.88	3.30	3.68	3.50
2nd	1.57	1.91 (1.31)	2.27	2.41	2.06
3rd	1.00	1.34 (0.91)	1.64	1.44	1.41
4th	0.70	1.00 (0.68)	1.20	1.00	1.13
5th	na	na	0.89	na	0.89
Final drive ratio, :1	2.95	3.07	4.06	2.47	3.44
Chassis & Body					
Layout	trans front engine/fwd	front engine/rwd	mid engine/rwd	front engine/rwd	rear engine/rwd
Body/frame	unit steel	fiberglass/skeletal steel	steel/skeletal steel	unit steel	unit steel
Steering type	rack & pinion, pwr-asst	rack & pinion, pwr-asst	rack & pinion	recirc ball, pwr-asst	rack & pinion
Steering overall ratio, :1	15.6	15.5	na	na	17.8
Brake system, f/r	10.2-in. vented discs/10.0-in. discs, ABS, hydraulic-asst	13.0-in. vented discs/12.0-in. vented discs, ABS, vacuum-asst	10.7-in. vented discs/10.9-in. vented discs, vacuum-asst	10.9-in. vented discs/11.0-in. vented discs, ABS, vacuum-asst	11.1-in. vented discs/11.4-in. vented discs, vacuum-asst
Wheels	alloy, 15 x 7JJ	alloy, 17 x 9½	alloy 16 x 7J	alloy 15 x 7J	alloy, 16 x 6J f/16 x 7J r
Tires, f/r	Goodyear Eagle VL, 225/60VR-15	Goodyear Eagle ZR40, 275/40ZR-17	Goodyear Eagle, VR-55 205/55VR-16/VR-50 225/50VR-16	Pirelli P6, 205/65VR-15	Goodyear Eagle, VR55 205/55VR-16/VR50 225/50VR-16
Suspension, f/r	MacPherson struts, lower A-arms, coil springs, tube shocks, anti-roll bar/Chapman struts, lower A-arms, transverse fiberglass leaf spring, tube shocks	upper & lower unequal-length A-arms, transverse fiberglass leaf spring, tube shocks, anti-roll bar/upper & lower trailing arms, lateral arms, tie rods, halfshafts, transverse fiberglass leaf spring, tube shocks, anti-roll bar	front & rear: unequal-length A-arms, coil springs, tube shocks, anti-roll bar	unequal-length A-arms, coil springs, tube shocks, anti-roll bar/ semi-trailing arms, coil springs, tube shocks, anti-roll bar, torque compensator	MacPherson struts, lower A-arms, torsion bars, tube shocks, anti-roll bar/semi-trailing arms, torsion bars, tube shocks, anti-roll bar
Accommodations					
Seating capacity	2	2	2	2	2+2
Head room, in.	36.0	36.5	34.5	35.5	37.5/31.0
Seat width	2 x 20.5	2 x 20.0	2 x 18.0	2 x 21.5	2 x 19.5/2 x 13.0
Trunk space, cu ft	14.1 + 4.0	6.0	5.3	9.4 + 5.8	5.2 + 4.1

[1] As-tested price includes for all five cars, std equip. (air cond, elect. window lifts, elect. adj mirrors); for the Allanté, std equip. (trip computer, ABS, hard top, elect. adj seats), Calif. emission controls ($99); for the Corvette, std equip. (ABS), leather seats ($1025), Z52 Sports Handling pkg ($970), AM/FM stereo/cassette ($773), elect. adj seats ($480), elect. a/c controls ($150), Calif. emission controls ($99), misc options ($35); for the 560SL, std equip. (ABS), West Coast base price differential ($565), gas guzzler tax ($1300); for the 911, std equip. (elect. adj hgt for driver seat, heated mirrors, elect. act. top ($2524), leather interior ($1444), fully elect. adj seats ($940), 16-in. forged wheels ($748), elect. a/c control ($603), driver lumbar support ($515), sport shock absorbers ($297), headlight washers ($232), carpeted luggage area ($168), AM/FM stereo/cassette upgrade ($153), misc options ($57)

PERFORMANCE

	Cadillac Allanté	Chevrolet Corvette Convertible	Ferrari 328 GTS	Mercedes-Benz 560SL	Porsche 911 Cabriolet
Acceleration					
Time to distance, sec					
0–100 ft	3.6	2.9	3.2	3.3	3.3
0–500 ft	9.3	8.0	8.4	8.4	8.3
0–1320 ft (¼ mi)	17.2	14.6	15.0	15.2	15.0
Speed at end of ¼ mi, mph	81.0	95.5	96.5	90.0	96.0
Time to speed, sec					
0–30 mph	3.2	1.8	2.5	2.7	2.5
0–60 mph	9.5	6.0	6.7	6.8	6.5
0–80 mph	16.9	10.0	10.4	11.5	10.8
Estimated top speed, mph	119	158	149	137	149
Fuel economy, mpg	17.1	19.3	18.2	16.3	22.5
Braking					
Stopping distance, ft, from					
60 mph	148	135	137	130	132
80 mph	257	225	234	231	237
Pedal effort for 0.5g stop, lb	17	18	22	na	33
Brake fade (six 0.5g stops from 60 mph), % increase in lb	12	nil	nil	na	9
Control	excellent	excellent	very good	excellent	very good
Overall brake rating	very good	excellent	excellent	excellent	excellent
Handling					
Lateral acceleration, g	0.77	0.87	0.87	0.78[1]	0.88
Slalom speed, mph	61.0	64.5	62.9	60.1[1]	64.5
Interior noise, with soft top/hard top, dBA					
Idle in neutral	49/47	54	67	46/46	56
Maximum, 1st gear	72/71	80	86	79/78	83
Constant 30 mph	62/56	66	72	58/56	68
50 mph	65/63	72	77	66/62	73
70 mph	71/68	76	78	76/68	78
90 mph	75/na	79	81	na	81

[1] From previous testing, November 1986.

CALCULATED DATA

	Cadillac Allanté	Chevrolet Corvette Convertible	Ferrari 328 GTS	Mercedes-Benz 560SL	Porsche 911 Cabriolet
Lb/bhp (test weight)	21.4	14.2	12.8	16.4	14.4
Bhp/liter	41.6	42.7	81.6	40.9	67.6
Engine rpm @ 60 mph in top gear	1750	1650	3100	2110	2590
R&T steering index	na	0.97	1.30	1.06	1.02

quently. Actually, the Allanté was already waxed when we got it. On the negative side were its relative lack of punch, questionable quality in things like the top mechanism and a feeling of fragility compared to others.

To summarize the Corvette as third choice in town and least favorite among wine country tourers doesn't do it justice. As is usual with our bunch, the Corvette positively polarized views. Two of our drivers put it second only to the 911 for daily use; others placed it bottom of both lists. Among its virtues were "torque and grip enough to dust off any wimps who come close" and being "an underrated bargain that'll outperform any other car in this bunch." However, noted a dissenter, "it rides like a truck and is seen in the company of the wrong people." Funny, because another of us said, "I guess I'm a Corvette kinda guy." And he's perfectly okay.

It was more than the lovely ambiance, the wonderful wine country setting and, yes, the wine itself. These are five delightfully distinctive convertibles, each with a unique personality, each offering something special.

Which is best? Maybe we better order up another bottle of wine and discuss this some more.

Should it be an elegant Chardonnay? Or would you prefer an exuberant Zinfandel? Maybe a classic Cabernet? Or would the subtlety of a Merlot be more to your taste? Or the undiscovered complexity of a Pinot Noir?

SUBJECTIVE RATINGS: WHAT'S YOUR CHOICE FOR

	The wine country	Daily use
Cadillac Allanté	Exclusive, easy to use; but lacks power and grip. Score: 11	Many creature comforts; but unknown quality. Score: 12
Chevrolet Corvette	Benign at limit; but harsh riding and feels large. Score: 8	Power and grip galore; but clunky feel. Score: 14
Ferrari 328 GTS	True exotic, all the right sounds; but limited space. Score: 23	Emotional lift; but will it ever be secure? Score: 9
Mercedes-Benz 560SL	Elegant, gobs of torque; but a bit floaty to some. Score: 11	3-pointed star, built to last; but dated style. Score: 16
Porsche 911	Fun at sub-limit, good power; but tricky handling. Score: 22	A useful exotic, sturdy feel; but odd ergonomics. Score: 24

DRIVING IMPRESSION

'89 NEW CARS
Cadillac Allanté

Responding with gusto.

BY PATRICK BEDARD

• The Allanté is a semi-important car for a lot of reasons: it pushed Cadillac into the over-$50,000 price stratosphere, it launched a co-production relationship with Pininfarina that opened the door to a world of possibilities, and it brought two-seater fun and frolic to a drab and lifeless model lineup. So much for the semi-important stuff.

The Allanté is *real* important for one reason: this will be the first post-energy-crisis Cadillac on which the company hasn't backed down from its vision just because of poor sales. Ford did the hard thing in 1984 when buyers stayed away in droves from its new international-style Lincoln Mark VII. Ford made the Mark VII more and more international until the right customers got the message. Now the car and the company are on a roll.

Cadillac, on the other hand, responded to slow sales of its de-baroqued 1986 Eldorado by troweling on the baroque again. That jump-started sales, but it also helped to keep Cadillac in the baroque-car business—a bleak place to be as the aging baroque enthusiasts slowly begin to trade down for wheelchairs.

The Allanté has earned a loser's reputation in the market, mostly because the company announced sales targets that no two-seat newcomer with an over-$50,000 price tag could hope to achieve. From the beginning, though, we've judged the Allanté itself to be a nice piece of work.

So we're particularly pleased that Cadillac resisted its usual phony-wire-wheel and padded-vinyl-roof impulses. Instead, the engineering department was allowed to make the Allanté more Allanté. The result, based on a drive in a pre-production example, is just flat wonderful.

If you like the concept—a two-seat con-

Vehicle type: front-engine, front-wheel-drive, 2-passenger, 2-door hard- and soft-top convertible
Estimated base price: $58,000
Engine type: V-8, aluminum block and iron heads, Cadillac electronic engine-control system with port fuel injection
Displacement . . . . 273 cu in, 4467cc
Power (SAE net) . . . . 200 bhp @ 4400 rpm
Transmission . . . . 4-speed automatic with lockup torque converter
Wheelbase . . . . 99.4 in
Length . . . . 178.6 in
Curb weight . . . . 3500 lb
EPA fuel economy, city driving . . . . 15 mpg

vertible tourer with hard and folding tops—the 1989 improvements will inspire you to get your grandmother *and* your left arm appraised.

Some of the changes are basic improvements. The soft top is much easier to erect now. The seats are more comfortable, and the leather is softer. The body's resistance to convertible quivers—which was very good before—is even better.

Some changes enhance controllability. The steering effort and the shock damping increase with speed now.

And some changes follow from the company's decision to let the fuel economy drop into gas-guzzler territory. "All of our competition is already there," said one engineer. "To match their performance, we had to give up some economy." But not much: the Allanté's EPA city and highway numbers are 15 mpg and 22 mpg, respectively.

Gained in the trade is 30 horsepower; last year's 4.1-liter V-8 is replaced by a high-output 4.5-liter that produces 200 hp. Adding to the Allanté's vigor are a new final-drive ratio (shortened from 2.95:1 to 3.21) and stickier tires, now on sixteen-inch wheels.

What the 1989 Allanté delivers for all this tinkering is refined vroom. Hey, drop the top. It takes two minutes, max. Light up the aluminum-block 4.5. Listen to the exhaust note, a V-8 hungry for pavement. Why resist?

The inside of a top-down Allanté is a good place to forget your troubles. The power is generous now, approaching thrust. The speed-variable damping removes all the Detroit queasies: the Allanté moves like a German. And the machine sends up subtle signals of quality design; the wind currents and open-car shakes are so well managed your author kept forgetting he was in a convertible. The Allanté's cockpit is a happy place to be.

In fact, your author decided—after blowing the dust off a piece of archaic imagery—this really was the Cadillac of two-seaters. •

Driving Impressions

CADILLAC ALLANTÉ

Keeping the faith... and the faithful

BY JOHN LAMM

THERE'S NO NEED to dwell on what a disappointment the Cadillac Allanté has been to many of us: It's been an even bigger one for the Cadillac/Pininfarina team that has been building it. Lagging sales, discounted prices and poor resale values are among the discouraging signs that make one wonder if Cadillac won't simply cut its losses and run. Executives say the question of the company's commitment to the car is another reason buyers have passed over the Allanté. Now Caddy hopes to win back their confidence with an upgraded car that's meant to assure them of the firm's dedication to the luxury ragtop.

Among the major changes to the Allanté for 1989 is the increase in the transverse V-8's displacement from 4.1 liters to 4.5. In addition, the powerplant gets new cylinder heads with bigger valves and other changes to the intake and exhaust ends that improve the engine's breathing. Compression ratio for the V-8 has been upped to 9.0:1. And it now has a nice, crisp exhaust note.

Cadillac says all this means that horsepower is up—from 170 bhp at 4300 rpm to an even 200 at 4400. Torque has been increased, too—from 230 lb.-ft. to 265 at 3200 rpm. With a change from a 2.95:1 final drive ratio to a 3.21:1, the Allanté can get to 60 mph in less than 8.5 seconds and reach a top speed in excess of 130 mph, according to the factory.

There's more—a Cadillac/Delco-developed "Speed Dependent Damping" system automatically sets the shocks for compliant, normal or firm ride depending on speed or rate of acceleration or braking. Other improvements include P225/55VR-16 Goodyear Eagle VL tires and rack-and-pinion steering that now has speed-variable assist to keep effort low for parking.

Interior changes are few and include new colors and softer leather upholstery. Anyone who has tried to raise or lower the soft top of an Allanté will be happy to know that it has been reworked with a new fabric and extra springs to make it much easier to operate.

Perhaps the fairest way to summarize all this is to say that the Allanté is definitely improved in a manner that shows Cadillac is really listening and reacting. The Allanté is quick, handles decently and is a delight to be in with the top down. Quality seems to be up to the level you'd expect from a car in this class.

Performance now fits the image. The Allanté feels much more responsive, and during some 90-plus-mph touring on two-lane roads, the car felt steady, secure and supple—as you'd expect for its price class and ambitions.

TOPLESS COMPARISON TEST

Cadillac Allanté versus Mercedes 560SL

The heavyweight two-seaters duke it out.

BY PATRICK BEDARD

- This is a title match, heavyweight two-seater division. The undisputed champion Mercedes-Benz SL is stepping into the ring yet again, one more time in its amazing eighteen-year-long career. This time the challenger is the brash young Cadillac Allanté. "Palooka," says the scuttlebutt. "The Allanté is all hype and no punch. Look at its record in the showrooms. This is gonna be another Mercedes KO."

For sure, the champ has built a career on knockouts. The contenders look good on paper, but they always come up short. What most ringsiders don't know, however, is how hard this challenger has been training for 1989. We previewed the toughened Allanté at a recent workout, and it showed us some impressive new moves.

Still, the SL's record is formidable—the envy of all the prestige-car makers. Sales results for 1988 are not complete as this is being written, but Mercedes sold 11,964 SLs stateside in 1987 and 12,530 in 1986. For a car that's always been priced in the stratosphere, that kind of showroom success is astounding.

That kind of success, it must be said, is also exactly what brings out the challengers. The 1989-model 560SL lists for $64,230. If past sales continue, that amounts to a nice little three-quarter-billion-dollar-a-year business. Would other carmakers like a bite of that? Do sharks like steak tartare?

We'll tell you how much one particular German maker wants a piece of the SL's action. Have you seen the recent print ad for the Porsche 928S4, the one with the headline that reads "Think of it as a Mercedes with Tabasco sauce"? Huh? Is there any way you can squint your mind's eye so that the fastback-sleek and enormously powerful 928 comes across as some kind of spicy Mercedes? We can't either. But if the champ were beating *you* up in the market—really hammering you despite your best offensive efforts—perhaps you'd blurt out whatever desperate thing came to mind, too.

How bad is the champ killing them in sales? Worse than you'd think. In 1987, Porsche sold 1967 928s, and 1988 is off about 30 percent from that. In a good Porsche year, Mercedes moves about six times as many SLs.

Jaguar does better in the high-buck two-seater business. In 1987, it sold 5380 XJ-Ss, and 1988 sales are close to the

ALLANTE VERSUS 560SL

same number. But the SL still outsells the Jaguar by more than two to one.

All of this evidence is being piled on the table not to show the weakness of the challengers—they are hardly unattractive machines in our estimation—but to dramatize the overwhelming success of the Mercedes-Benz 560SL. This car was introduced to the U.S. market in 1972 as the 350SL, priced at $10,500 (that was a stiff ticket in those days, but in constant dollars the car is far more expensive today). Since then it has been face-lifted, refined, and upgraded, but it's still easily recognizable as an evolution of the 1972 model. Unlike most cars, which slow in sales as they age, the SL has trended upward to today's awesome total. The Mercedes-Benz SL convertible now defines the class. When seriously affluent Americans think of a chic sporter for two, most of them think of this car.

And if the customers regard the SL as the standard, then so must the challengers (and so must we). The Allanté was created with the SL in mind. Cadillac never said that exactly, but it did point to the customers it wanted—and they were all heading for their local M-B stores. How could they be intercepted? Showing them a better SL was the obvious strategy. Cadillac held nothing back with the Allanté. It even confronted its own Not Invented Here Syndrome, admitted that outside help was necessary, and hired Pininfarina to design and manufacture the bodies and help with general development. Cadillac, you must understand, didn't just roll another model out the door: the Allanté is a premeditated and carefully executed move to sock it to the champion SL.

How's it going so far? Not badly, we'd say—despite what you may have heard about weak punches in the showroom. In model-year 1987, the introductory year, Cadillac reported 1651 Allanté sales, followed by 3502 sales of the 1988s. Admittedly, this is not heavy traffic and it didn't cause much sweat at Daimler-Benz.

Nonetheless, we think the Allanté made a respectable showing. We say that for two reasons: (1) its over-$50,000 introductory price was twenty grand more than anything else wearing the Cadillac label, which means there were no customers predisposed to over-$50,000 Cadillacs, and (2) the Cadillac label gets no respect in the ultrahigh end of the business. Moving the iron against that sort of head wind is necessarily slow going.

But so what? Champs aren't made in a year or two. Look at the SL. It has secured its lock on the market over eighteen years. There's only one question: Is the Allanté a better car? If it is, the sales will come in due time.

Is the Allanté a better car? That is the question, isn't it? And that's the reason for this title match. Why wait years for the vote to come in when a few days of driving and a few trips to the test track can serve up the answer immediately?

The Weigh-In

At the weigh-in, the SL and the Allanté showed the similarities you'd expect from closely matched opponents. In external dimensions the Allanté is chunkier, about two inches shorter and three inches wider, and stands on the road with a widespread stance: its wheelbase and track are each nearly three inches greater than the SL's.

At 3681 pounds, the SL outweighs the Allanté by 163 pounds. It also has a far larger engine: 5547cc, versus 4467cc for the Allanté. Horsepower is 227 for the Mercedes, 200 for the Cadillac. Both cars are two-seater convertibles with removable hard tops. Neither is suitable for a third occupant behind the buckets; indeed, riding back there would be virtually impossible in the Allanté.

Both are powered by V-8 engines coupled to four-speed automatics, but here the first significant difference appears. The SL is rear-drive; the Allanté's V-8 powers the front wheels.

Both have independent suspension and disc brakes at all four corners, and both have anti-lock brakes as standard equipment.

Round One: First Impressions

Boy, you can really tell the champ from the challenger. It's old versus new. The SL has an instrument panel full of white-on-black gauges positioned high on the dash where you can get a close look at them. The Allanté has a glowing all-electronic cluster—arranged in ersatz dials and bar graphs—that shows you a palette of red, blue, green, gray, and yellow. For traditionalists, a proper needles-and-numbers cluster is available on the Allanté at no charge, but Cadillac has always maintained that the luminous electro display was the right way to monitor this car's soul.

The different approaches to instrumentation provide the first indication of an age difference between the two cars, but there are plenty of others. The SL, amazingly, still has bright-metal (instead of black) windshield-wiper arms and a Florsheim-shiny material stretched across the top of the dash. Both reflect sunlight excessively. The SL's cockpit is notably chromey, and the grain of its upholstery, particularly on the door panels and the console, has a bygone-years aura about it. You also see plenty of fine-furniture-quality wood trim.

There's no wood in the Cadillac. Instead, it greets you with a "bink, bink" key warning and a "plink, plink" turn-signal indicator, both electronic synthesizer sounds. Your eye falls on smoothly molded shapes with non-glare finishes. The dashboard buttons—dozens of them—are systematically arranged in rows and columns, and there are power assists for everything, including the complex seat adjustments.

The SL is a stark contrast, still reflecting the thinking of two decades ago when power seats weren't necessary, electronic displays didn't exist, and Mercedes decided that a power-adjustable mirror was something installed only on the right side—where the driver couldn't reach a mechanical adjuster.

Round Two: On the Road

The SL's cockpit is that of a traditional two-seater, which is to say it's intimate. You sit near the floor, and everything around you is sports-car close, including the dash and the windshield.

The wider Allanté, on the other hand, is cavernous, a room with two chairs in it. The windshield is well forward and the doors are so far away that the armrests are nearly unusable.

The cars respond differently to your touch, too. The M-B feels a bit ponderous. The throttle moves a long way before anything happens. You get used to that but not to the slowness of the transmission's downshift decisions. Full-throttle kickdowns at city-traffic speeds are a matter of much deliberation—and of too much lunge.

The Allanté is quick and crisp in traffic. Its engine is somewhat louder, mostly exhaust actually, and the note is okay if you like that kind of thing. The M-B sounds more refined and quieter at low speeds, changing when the revs top 4500 to the hard metallic song of purposeful machinery. This car has subtle ways of reminding you that something pretty wonderful resides under the hood.

Because these are convertibles, we made a point of checking them for the body shakes and quivers inherent in topless construction. Daimler-Benz pioneered the use of special latches that trans-

ALLANTE VERSUS 560SL

form the doors, when closed, into structural panels. For the same reason, notably solid clamps attach the SL's hard top. This car was a marvel of convertible rigidity when it was new in 1972.

Our comparisons focused on soft-top and top-down driving because these are the worst conditions for shake. The jittery feeling imparted by Los Angeles freeways seemed a bit harsher in the SL, while the Allanté showed more windshield shake. At cruising speeds with their soft tops up, you hear wind roar in the SL, exhaust roar in the Allanté. While the cars were definitely different in the way they responded to road inputs, we judged them to be equal—and very acceptable—in their annoyances.

Round Three: Convertibility

These are casual sporters, not high-performance GT missiles, and top-down motoring is a big part of their attraction. So how are they as convertibles?

One way to judge is to ask, How easily do the tops go up and down? (Neither is power-assisted.) This question is complicated, of course, by what level of closed-car perfection each car is seeking when the top is up. The Allanté has a hard glass rear window with integral de-icing. The SL has a plain old plastic-film window complete with vision-distorting waves when you look through the inside mirror; in other words, a little behind the times even for 1972.

Having said that, the SL's soft top was easier to put up and down, even though it required the use of several special tools. The Allanté took more tugging and grunting to make latches meet, likely because the car was so young (only 52 miles showing when delivered) that the fabric had not yet taken its stretch. Probably once you've been through the procedure a dozen times to learn the quirks, the two are about the same.

For hard-top removal, however, they'll never be equal. The M-B's top weighs a hefty 96 pounds, versus 58 pounds for the Allanté's. In the SL's handbook there is the polite suggestion that maybe this is a job for the dealer. For some owners, it surely would be.

A second way to judge the joys of convertibility is to ask, How drafty is the cockpit with the top down? Here the Allanté is vastly superior. Pininfarina's aerodynamic tweaking paid off. In what is admittedly a highly subjective comparison, we concluded that the "windiness" level felt in the Allanté at 60 mph was equaled by the SL at 47 mph, give or take a little. The Allanté delivers fully on its open-air promise.

Round Four: The Mountain Road

Racetrack performance is not the point of these casual sporters. Far more important to ask is, How are they on that mythical mountain road? The Allanté has new tricks for 1989 aimed at just this sort of driving: sixteen-inch wheels with Goodyear Eagle VL 225/55VR-16 tires, power steering that increases in effort as speed rises, and shock absorbers that automatically (no cockpit switches) increase damping with speed. In addition, the shocks stiffen between 0 and 5 mph to reduce front-end lift during accelera-

C/D Test Results

	acceleration, sec					top speed, mph	braking, 70–0 mph, ft	roadholding, 300-ft skidpad,
	0–60 mph	0–100 mph	¼-mile	top gear, 30–50 mph	top gear, 50–70 mph			
CADILLAC ALLANTÉ	8.4	28.9	16.5 @ 82 mph	4.1	5.7	127	186	0.80
MERCEDES-BENZ 560SL	7.1	19.8	15.6 @ 90 mph	3.6	5.0	136	178	0.79

Vital Statistics

	price, base/as tested	engine	SAE net power/torque	dimensions, in			
				wheel-base	length	width	height
CADILLAC ALLANTÉ	$56,533/$57,183	V-8, 273 cu in (4467cc), aluminum block and iron heads, Cadillac electronic engine-control system with port fuel injection	200 bhp @ 4400 rpm/ 270 lb-ft @ 3200 rpm	99.4	178.6	73.5	52.2
MERCEDES-BENZ 560SL	$64,230/$65,780	SOHC V-8, 338 cu in (5547cc), aluminum block and heads, Bosch KE-III-Jetronic port fuel injection	227 bhp @ 4750 rpm/ 279 lb-ft @ 3250 rpm	96.7	180.3	70.5	51.1

tion; and, to limit dive, they firm up with any brake application above 35 mph.

The result of these chassis advancements is a profound change in the demeanor of the Allanté. The quick-reflex tires and always-right shocks transmit a tight, in-touch-with-the-road sensation that used to be pretty much the exclusive property of German cars.

The SL, with its more conventional underpinnings, feels like the traditional Detroiter. Its ride is less harsh. Its body rolls more in the turns. And its shocks feel floaty over the bumps.

In the twisties, the Allanté has crisper steering response and the self-firming shocks do a very nice thing—they resist that nose-down, tail-up pitch that comes when you pull back on the power as you pick up the turn's arc. As you rush into the turns, the Allanté feels secure. In the SL, you feel the tail rise, then tip to the outside, followed momentarily by a side step of the rear tires. It's a bit messy.

The Allanté got messy on one particular bump, which set its rear wheels to dancing. We tried this disturbance a number of times in both cars, and the SL barely noticed.

Although front drive is usually not the preferred choice for sporting cars, the front-drive nature of the Allanté is apparent only in low-speed turns—intersections, for example—when you're accelerating hard. Out on the twisties, at speeds appropriate to these casual runabouts, it was undetectable.

These speeds did reveal different approaches to seat design, however. The SL has typical M-B firm-but-flat cushions topped by a backrest that's curved, as you look down on it, to cup you in place. The Allanté has a cushion with high side bolsters and a relatively flat backrest. In effect, the SL tries to keep your upper body from sliding sideways; the Allanté tries to hold your butt in place. The SL's seat works less well, we think, but the door is so close that it's easy to brace against it. The door is too far away in the Allanté, which means you brace your upper body against the wheel. Opinion is divided about which system works best.

Round Five: Pedal to the Metal

The drag race was inevitable. Line 'em up and let it happen. The big-engined SL, pulling fiercely to its 6000-rpm redline, made itself small in the Allanté's windshield. No surprise.

What was surprising, however, was the opening stages of the race. The Allanté has gained 30 horsepower for 1989 from a much-revised engine (an extra 380cc is just a small part the package). Off the mark it opens up a car-length lead and holds it until the M-B is about halfway through second gear. Then the German's superior power takes over and the race quickly becomes no contest.

The M-B wins if you hold the pedal down long enough, but in traffic the Allanté is quicker. Its shrewdly chosen ratios and its quick-to-downshift transmission use the smaller engine to best advantage. The fact that the M-B hesitates so long before reaching for a lower gear means that, once the cars are rolling, the Allanté will always win the jump. And at urban speeds, it stays ahead.

The Decision: It's Unanimous

The judges say no KO. But they are unanimous in their decision. The victory goes to—ta-dah!—the Allanté.

The SL is powerful, it's nicely screwed together, and it oozes prestige. But it's old. Its reflexes are slow. Its cockpit lacks comfort (compared with the Allanté's), and driving it seems more of a burden.

The Allanté is unfailingly hospitable and feels delightfully quick to the touch in normal driving. It ends up being more likable.

Daimler-Benz has not been asleep, of course. For years, spyphotos of a new SL have been slipping out of Stuttgart. The company now confirms that the replacement will be here in time for the 1990 model year.

For now, though, it's time to update the tip sheet: this new kid, the Allanté, has been hugely underrated. ●

road horsepower @ 50 mph	interior sound level, dBA				fuel economy, mpg		
	idle	full throttle	70-mph cruising	70-mph coasting	EPA city	EPA highway	C/D observed
17	45	75	70	70	15	23	12
15	44	81	71	71	14	17	13

curb weight, lb	weight distribution, % F/R	suspension		brakes, F/R	tires
		front	rear		
3518	61.3/38.7	ind, strut located by a control arm, coil springs, anti-roll bar	ind, strut located by a control arm, transverse leaf spring	vented disc/disc; anti-lock control	Goodyear Eagle VL, P225/55VR-16
3681	51.1/48.9	ind, unequal-length control arms, coil springs, anti-roll bar	ind, semi-trailing arm, coil springs, anti-roll bar	vented disc/disc; anti-lock control	Pirelli P6, 205/65VR-15

San Pedro, California—

It seems like such a good idea, the kind of thing we're always proposing over lunch to assorted members of Detroit's power elite. Ship a bunch of car bits to a famous Italian manufacturer, let said manufacturer wrap a slick sheet-metal shape around them, and then bring the whole business home and install the powertrain. Finally, you sell the complete car for a ton of money and retire with your bag of loot to Lake Charlevoix.

But if this is such a good idea, why aren't we rich and famous? More important, why isn't the Cadillac Allanté a wildfire success? These questions keep us awake at night, and we're hoping to put them to rest by taking a 1989 Cadillac Allanté into our Four Seasons fleet. After all, we feel as if this car was practically our idea to begin with.

It is no secret that the Allanté has had a troubled time in the marketplace. From the time of the car's introduction in 1987, traditional Cadillac buyers appeared to be scandalized by the price (currently some $57,000), and import buyers were unpersuaded by the performance. As a consequence, sales have never come close to the initial projection of 8000 cars per year. Only 1651 were sold in 1987, 3502 in 1988, and probably fewer than 3200 this year. Cadillac halted production for two months last year yet still finished up with a 262-day supply of the car. Pininfarina has even begun to solicit other projects to take up the slack on the Allanté production line in Turin.

Still, the Allanté's troubles might have more to do with the marketplace than with the car itself. The field of luxury two-seaters has grown extremely crowded lately, and buyers are becoming price-sensitive. Sales of the $30,000 Buick Reatta plateaued soon after its successful 1988 introduction, and a new

FOUR SEASONS TEST

CADILLAC ALLANTÉ

It's already the standard for "Dallas." When will this car become the standard for the world?

BY MICHAEL JORDAN
PHOTOGRAPHY BY VIC HUBER

advertising campaign has been geared up to boost sales to 8000 cars. And although the $33,000 Chrysler's TC by Maserati has apparently sold well at the dealer level, it is too soon to gauge whether the car will reach its target of 3500 sales this year. Meanwhile, sales of the Porsche 928S4 have declined (to 1700 in 1988), although the Mercedes-Benz 560SL's have held steady (at 11,500 in 1988).

But there's more to this car than gossip about sales numbers. Such talk tends to distract people from the worthwhile increase in the numbers produced under the hood this year. Cadillac might have made it perfectly clear at the Allanté's introduction that this was a luxury car, but the engineers have been dragging the project toward the sporting end of the market ever since.

To begin with, the V-8 engine has received a useful 4mm increase in bore size, increasing displacement nine percent to 4.5 liters. Then Jack Roush's racing engineers reportedly went to work on the package, helping to produce new cylinder heads with more efficient porting and bigger valves, a taller 9.0:1 compression ratio, and less restrictive exhaust plumbing. There is a whole laundry list of upgrades for the fuel-injection system and the transmission besides. The bottom line is an eighteen percent increase in power to 200 bhp at 4300 rpm, and a thirteen percent in-

crease in torque to 270 pounds-feet at 3200 rpm. These are serious numbers, especially in conjunction with a final-drive ratio that is nine percent shorter. Cadillac claims the 1989 Allanté will reach 60 mph 1.5 seconds quicker (8.5 versus 10.0) than the version we tested in September 1986.

The chassis people haven't been sitting around with their feet propped on their desks, either. The shock absorbers now have three-way, speed-sensitive damping control, which automatically comes into play at thresholds of 25 mph and 60 mph and also counters squat and dive during acceleration and braking. Speed-sensitive, variable-assist power steering is new for the 1989 Allanté as well. Finally, the Goodyear Eagle VL tires (a quiet-ride version of the Eagle GT) are more aggressive this year, thanks to a lower 55-series profile.

Frankly, we think the Cadillac guys ought to ditch all their euphemisms and start calling the Allanté a sports car. A friend of ours endorsed that view during a drive from Los Angeles to Las Vegas and back. Since he makes his living by calibrating the speedometers of civilians (with appropriate monetary penalties for wrong answers), our friend was impressed with the Allanté's minimal 0.5-mph speedometer error at 80 mph, not to mention the car's 134-mph (indicated) capability. He was even more impressed with the Allanté's four-wheel disc brakes and Bosch III ABS. He said later, "I got into the brakes at 134 mph, and I expected them to be wimpy and the pedal to start vibrating with the ABS right away. But the pedal stayed nice and firm, and I kept pressing harder and harder, and it kept stopping. By the time I got to 20 mph, I was practically dizzy from the deceleration. Great brakes."

Our man didn't have much luck profiling in Las Vegas, however. As he noted in our logbook, "When I pulled up next to a Mercedes-Benz 560SL, its driver turned to look, pointed, said something to his passenger, and they both laughed." We're inclined to think maybe it was the driver they were pointing at, because everyone else has had great luck with the Allanté's crisply drawn Pininfarina creases. While at the wheel of our Allanté, business manager Harriet Stemberger captivated the doorman at the Amway Grand Plaza hotel in Grand Rapids and then later mesmerized her son's college roommates. She noted in our logbook, "I don't think I've ever been in a car that attracts so much attention. To think that one small red car could attract a whole table of college football players. Be still, my heart!" Even in the depths of jaded Southern California where the Allanté continues to be a rarity, our people were accosted at carwashes and followed by carloads of giggling young women.

This status confrontation with cars like the Mercedes-Benz 560SL is actually no joke. The Allanté was expressly designed to put Cadillac smack-dab in the middle of the automotive enthusiasms of the super-rich, among cars like the 560SL, the BMW 635CSi, the Jaguar XJ-S, and the Porsche 928S4. At the same time, the Allanté was meant to interpret this idiom for the mainstream Cadillac owner. That is why we were very interested in the reaction of John Phillips, Jr., who drove our test car from Michigan to California. Phillips is a sixty-seven-year-old attorney from Columbus, Ohio. Aside from an inexplicable enthusiasm for Cadillac limousines (he owns one and drives it just for fun) and his equally inexplicable parental relationship with John Phillips III, our extraterrestrial senior editor, Mr. Phillips fits the Allanté's demographics perfectly. Phillips and his wife, Jeanne, were wildly enthusiastic after their drive, completely taken aback by the crowds that formed around this car at every gas stop. They became so distracted that they even left their gas card behind at a station in Illinois (thanks for mailing it back, Darlene). Phillips later reported several bursts of triple-digit velocity, something he is unaccustomed to in his daily car, a 1980 Volkswagen Cabriolet.

Although we're only beginning our test, the Allanté's logbook is already brimming with comments, clear evidence that this car arouses powerful feelings wherever it goes. All of our impressions haven't been happy ones, though. The throttle linkage continues to

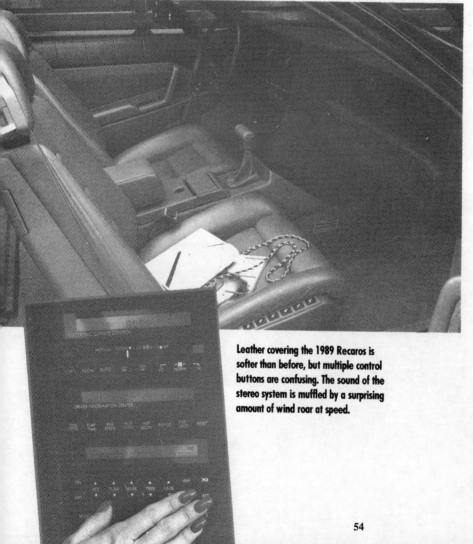

Leather covering the 1989 Recaros is softer than before, but multiple control buttons are confusing. The sound of the stereo system is muffled by a surprising amount of wind roar at speed.

furnish a curious nonlinear action, accentuated by the shorter gearing and horsepower boost. As John Phillips III notes in the log, "There are two settings: The first inch of travel feels like idle; every inch thereafter feels like unintended acceleration." Meanwhile, the cabin fills with a surprising amount of wind roar at speed, muffling the effect of an awfully nice stereo system. The control buttons that swarm alongside the power seats are impossible to learn, making the stony seat bolsters even more annoying (although the leather covering the 1989 Recaros *is* softer than before). The convertible top must be slam-dunked in place to keep it from suddenly giving way at speed and becoming some kind of massive fabric air brake. And the optional analog instrumentation, new for 1989, is woefully bland.

Like so many cars from General Motors, the Allanté has improved significantly since its disappointing introduction. The question remains, however, whether it can find its audience before the accountants lose patience with the program. As much as we enjoy this car, it still leaves us with more questions than answers. Is this a sports car? Does it have the quality to compete in its market? What do you really get for nearly $60,000? Is this a real American car or a marketing gambit? What will happen when Mercedes introduces its new SL (and then a V-12 upgrade) and BMW announces its new 8-series coupe?

In larger terms, the question here is whether Cadillac can really slug it out with BMW, Jaguar, and Mercedes-Benz to meet the goals that General Motors set for it in the massive corporate reorganization several years ago. We're looking forward to learning the answer.

The Allanté's Pininfarina styling creates a great deal of attention.

CADILLAC ALLANTÉ

GENERAL:
Front-engine, front-wheel-drive convertible
2-passenger, 2-door steel and aluminum body
Base price/price as tested $56,533/$57,183

MAJOR EQUIPMENT:
Removable hardtop standard
Air conditioning standard
AM/FM/cassette standard
Leather interior standard
Cruise control standard

ENGINE:
OHV V-8, aluminum block, iron heads
Bore x stroke 3.62 x 3.31 in (92.0 x 84.0mm)
Displacement 273 cu in (4467cc)
Compression ratio 9.0:1
Fuel system electronic multipoint injection
Power SAE net 200 bhp @ 4300 rpm
Torque SAE net 270 lb-ft @ 3200 rpm
Redline 5200 rpm

DRIVETRAIN:
4-speed automatic transmission
Gear ratios (I) 2.92 (II) 1.57 (III) 1.00 (IV) 0.70
Final-drive ratio 3.21:1

MEASUREMENTS:
Wheelbase 99.4 in
Track front/rear 60.4/60.4 in
Length 178.6 in
Width 73.5 in
Curb weight 3492 lb
Weight distribution 62/38%
Fuel capacity 22.0 gal

SUSPENSION:
Independent front, with damper struts, lateral and trailing links, coil springs, anti-roll bar
Independent rear, with damper struts, control arms, transverse plastic leaf spring

STEERING:
Rack-and-pinion, variable power-assisted

BRAKES:
10.3-in vented discs front
10.0-in discs rear
Anti-lock system

WHEELS AND TIRES:
16 x 7.0-in forged aluminum wheels
225/55VR-16 Goodyear Eagle VL tires

PERFORMANCE (manufacturer's data):
0–60 mph in 8.5 sec
Standing ¼-mile in 16.6 sec @ 83 mph
Top speed 130 mph
EPA city driving 15 mpg

MAINTENANCE:
Headlamp unit $80.00
Front quarter-panel $309.00
Brake pads front wheels $75.00
Air filter $17.45
Oil filter $8.00
Recommended oil change interval variable (computer-calculated)

	EXCELLENT	GOOD	FAIR	POOR
ENGINE				
power			•	
response			•	
smoothness				•
DRIVETRAIN				
shift action			•	
power delivery			•	
STEERING				
effort			•	
response			•	
feel			•	
RIDE				
general comfort		•		
roll control			•	
pitch control			•	
HANDLING				
directional stability			•	
predictability				•
maneuverability			•	
BRAKES				
response				•
modulation			•	
effectiveness				•
GENERAL				
ergonomics			•	
instrumentation			•	
roominess			•	
seating comfort			•	
fit and finish				•
storage space			•	
OVERALL				
dollar value			•	
fun to drive				•

LONG-TERM TEST

Cadillac Allanté
by Jim Miller

Louie, this could be the beginning of a beautiful friendship

PHOTOGRAPHY BY MIKE BANKS

Cadillac's Allanté has always been a head-turner. From day one, its understated elegance has struck just the right chord with passersby. But the first of these 2-seat beauties was a bit like a Hollywood set; there just wasn't much behind the façade.

In the past two years, the diminutive Caddie has seen major changes, however, and it's gone a long way toward putting some gold behind the glitter. Last year's modifications converted the Allanté into a real challenger in the luxury 2-seat market. Moreover, Mercedes' move *way* upscale with the SLs makes the Allanté—at $20,000-30,000 less— seem like a real bargain.

But it's earned its place by more than default. Along with other improvements, the '89 model year brought a major step up in performance. Cadillac replaced the wimpy 4.1-liter engine with the 4.5-liter V-8 the car deserved. In addition to increasing displacement, the division also boosted valve sizes, straightened the intake ports, and reworked the exhaust manifolds, catalytic converters, and mufflers for less restriction. Power jumped 17%, from 170 hp to an even 200, and torque climbed from 240 lb-ft to 270.

Suddenly there was muscle behind those svelte Pininfarina-sculpted flanks. Flexing that muscle with a firm push on the right pedal produces some most untraditional Cadillac responses. The exhaust note changes from a pleasant burble to a serious snarl, and the Allanté charges forward with élan. For the past several years, Cadillacs have provided some of the best engine and transmission pairings in the domestic market, and this top-of-the-line roadster upholds that trend. Transmission response is simply superb. The 4-speed auto snaps from one gear to the next without delay, and changes speeds faster

than a carnival barker.

And this year's driveline brings even better news. Combined with last year's horsepower infusion, the addition of Traction-Control means the Allanté also has the poise to know when to use that muscle. As a result, this Cadillac ranks with the world's best in both performance and sophistication.

For those who doubt such a claim, we'd suggest a quick trip through the gears on spotty pavement. When the traction matches the Cadillac's thrust, the 3466-lb convertible snaps from 0-60 in 7.9 sec. When the grip isn't quite up to the car's 270 lb-ft of torque, the Traction-Control steps in to manage the forward momentum by keeping the engine and tires on equal footing (see sidebar). Only a well-snubbed chirp from the tires and a quick ratcheting noise betray the system's existence. A TRAC-CTRL indicator lamp on the dash also gives the secret away. During typical dry-pavement driving, fast, hard turns are most likely to engage the system. Under those conditions, it not only eliminates any tire-smoking pyrotechnics, but effectively stifles torque

The Allanté falls under the heading of the best of Detroit

steer as well. When you're dealing with 200 hp and a set of sticky, 225/55VR16 Goodyears, that's a welcome side effect.

In general, the only messages you get through the steering wheel are those you want to receive. Last year, Cadillac endowed the Allanté with variable-rate power assist, and the steering consistently maintains a perfect balance between feel and effort. And unlike some other variable-assist designs, the rate increases gradually, so it never intrudes on the driver's awareness.

Likewise, the Speed Dependent Damping is completely transparent

to the driver, though the arrangement has been recalibrated since its '89 introduction. The deflected-disc adjustable shocks now remain in the softest, comfort setting below 40 mph, stiffen to normal from 40 up to 60, and switch to firm above 60 mph. The system also dials up the firm setting during braking or acceleration to reduce dive and squat. But just as they are with the steering, the transitions go completely unnoticed from the cockpit. Moreover, the suspension settings match the car's character perfectly. Cruising Lake Shore Drive or cutting along Angeles Crest Highway, the Caddie seems well-damped at any speed, yet the ride never becomes jarring. Understand that these aren't the traditional Chris-Craft ride motions. Like the Seville Touring Sedan, the Allanté manages to tread that fine line between athletic ability and black-tie elegance—like Greg Lemond in a tux.

The interior follows that balance. The monochromatic charcoal gray color scheme and stark, white-on-black gauges (digital is still optional) give our test car a serious, almost Teutonic appearance. But the interior is far more American in its treatment of the driver and passenger. The supple leather bucket seats offer a wide range of adjustment, and the driver's side features two memory settings. A full range of convenience features—from automatic headlamps and climate control to a trip computer and 170-watt Bose stereo/CD player—provide the expected Detroit opulence.

Yet, even with all the trappings, the dash remains exceptionally simple to decipher. The center console groups the 44 buttons for the climate control, trip computer, AM/FM/cassette stereo and CD player, but the displays break up the controls so that the console seems neither cluttered nor confusing. The easily read labels and logical layout make it simple to keep track of the Allanté's functions, despite their potentially bewildering number.

If the Allanté's luxury fittings fall short in any area, it's in the design of the top mechanism. The manual softtop is still more fiddly than it needs to be, particularly compared to the power tops offered by Porsche or Mercedes. Granted, those also come with a severe price penalty, but the five-minute exercise required to raise or lower the roof is a bit much.

Even more maddening, the mechanism that cinches the top seems extremely temperamental, so much so that metal expansion after several hours in the desert apparently kept the hook from catching the motor-driven latch. Ten minutes of pushing and a ballpark adjustment were required to get the top cinched down. When you consider that Buick's Reat-

> **Traction-Control means the Allanté also has the poise to know when to use its muscle**

Traction-Control—Turning the Tables on ABS

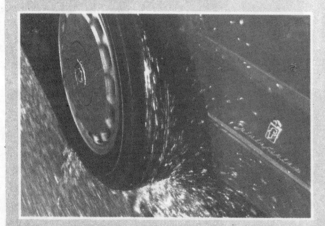 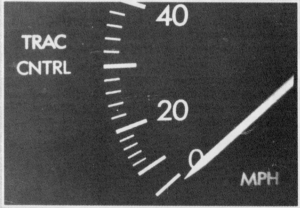

Slipped in among the Twilight Sentinel, cruise control, climate control, compact disc player, and the remaining electronic labor-saving devices in the '90 Allanté lurks the latest in dynamic car control. Cadillac's Traction-Control system, developed in conjunction with Bosch electronics, eliminates one more opportunity for driver error. By effectively limiting the amount of horsepower available to the driving wheels, the new Allanté stops wheelspin during acceleration.

Although the effects of the system are new, most of the hardware has been in place since the car's introduction. Basically, the Traction-Control combines functions of the anti-lock brakes and the engine control module (ECM). But the system works somewhat in reverse from ABS. Instead of selectively releasing the brakes, the Traction-Control applies one or both front calipers, depending on input from the ECM. A pair of speed sensors, one at each front wheel, signals the ABS control module whenever the tire starts to spin. Initially, the ABS computer pumps the brake on the slipping wheel, and continues to pulse so that the car maintains about 10% slip at each wheel until it is able to regain traction.

The setup uses more than just the brakes for control, however. Continuous slipping—such as on ice—causes the system to cut off fuel to the engine, one cylinder at a time. After 3 sec, the ECM drops one more, then another every 3 sec until traction is regained or the engine has been cut to four cylinders. This arrangement not only reduces stress on the engine, but the loss of power also signals the driver that the system is operating. After 2 sec, a yellow lamp on the instrument panel flickers on as a further warning. An additional system safeguard disconnects the Traction-Control whenever the small-diameter spare tire is removed from the trunk.

Like anti-lock brakes, the system is designed for worst-case scenarios—driving on ice, snow, rain—and has little effect on everyday operation. It's yet one more example of better driving through electricity. —*J.M.*

TECH DATA
SPECIFICATIONS & PERFORMANCE

Cadillac Allanté

GENERAL
Make and model	Cadillac Allanté
Manufacturer	Cadillac Motor Car Div., Detroit, Mich.
Body style	2-door, 2-passenger
Drivetrain layout	Front engine, front drive
Base price	$50,900
Price as tested	$51,650
Options included	Gas-guzzler tax, $650; California emissions, $100
Typical market competition	Mercedes 300 SL, Chrysler's TC by Maserati

DIMENSIONS
Wheelbase, in./mm	99.4/2525
Track, f/r, in./mm	60.4/60.4/1533/1533
Length, in./mm	178.7/4539
Width, in./mm	73.5/1866
Height, in./mm	52.2/1327
Ground clearance, in./mm	6.2/158
Manufacturer's curb weight, lb	3466
Weight distribution, f/r, %	56/44
Cargo capacity, cu ft	16.3
Fuel capacity, gal	22.0
Power/weight ratio, lb/hp	17.3

PERFORMANCE AND TEST DATA
Acceleration, sec	
0-30 mph	2.8
0-40 mph	4.2
0-50 mph	5.9
0-60 mph	7.9
0-70 mph	10.4
0-80 mph	13.7
Standing quarter mile, sec @ mph	16.1 @ 86.5
Braking, ft	
30-0 mph	31
60-0 mph	125
Handling	
Lateral acceleration, g	0.85
Speed through 600-ft slalom, mph	58.9
Speedometer error, mph	
Indicated	Actual
30	30
40	40
50	50
60	60
Interior noise, dBA	
Idling in neutral	50
Steady 60 mph in top gear	68

FUEL ECONOMY
EPA, city/hwy., mpg	15/22
Est. range, city/hwy., miles	330/484

ENGINE
Type	V-8, liquid cooled, cast alloy block, cast iron heads
Bore x stroke, in./mm	3.62 x 3.31/92.0 x 84.0
Displacement, ci/cc	273/4467
Compression ratio	9.0:1
Valve gear	OHV, 2 valves/cylinder
Fuel/induction system	Multi-point EFI
Horsepower, hp @ rpm, SAE net	200 @ 4400
Torque, lb/ft @ rpm, SAE net	270 @ 3200
Horsepower/liter	44.8
Redline, rpm	5200
Recommended fuel	Unleaded premium

DRIVELINE
Transmission type	4-speed auto.
Gear ratios (1st)	2.92:1
(2nd)	1.57:1
(3rd)	1.00:1
(4th)	0.70:1
Axle ratio	3.21:1
Final drive ratio	2.25:1
Engine rpm, 60 mph in top gear	1950

CHASSIS
Suspension	
Front	MacPherson struts, lower control arms, coil springs, anti-roll bar
Rear	Multi link, transverse leaf spring
Steering	
Type	Rack and pinion, power assist
Ratio	15.6:1
Turns, lock to lock	2.4
Turning circle, ft	40.2
Brakes	
Front, type/dia., in	Vented discs/10.3
Rear, type/dia., in	Discs/10.0
Anti-lock	Standard
Wheels and tires	
Wheel size, in	16 x 7.0
Wheel type/material	Forged alloy
Tire size	225/55VR16
Tire mfr. and model	Goodyear Eagle VL

INSTRUMENTATION
Instruments	145-mph speedo; 6000-rpm tach; fuel level; oil press; coolant temp; volts; digital clock
Warning lamps	Oil press; coolant temp; stop engine; service engine; emergency brake; belts; theft deterrent; trac control; check info center; passive restraint

ta offers a more elegant solution for more than $15,000 less, the Allanté top seems even more irksome. Cadillac has promised a reworked mechanism, due sometime early this year, which we'll report on as soon as we've seen it.

At least some of this frustration comes a bit cheaper this year. For 1990, Cadillac offers a soft-top-only version, after research showed that a large number of owners removed the hardtop once—to place it in storage. Consequently, the division now al-

lows buyers to eliminate the hardtop and save themselves $6000 in the process.

In the meantime, we're planning to keep this car around for a while. Along with a select few cars, the Allanté falls under the heading of the best of Detroit, and we're not willing to let it go without a thorough wringing out. The fact that it's rapidly become one of the most popular cars in the test fleet has something to do with our decision, of course. But we also want to see if the goodness is the real thing or just an illusion. **MT**

Second Opinion

The '90 Allanté is an excellent car that offers all a model in its class should offer—distinctive styling, wonderful creature comforts, and an un-Cadillac amount of performance and handling. Today's Allanté is a car the enthusiast driver will enjoy, yet it won't turn off the luxury car buyer with a harsh ride and noisy engine. The problem is, like many General Motors cars of recent memory, this excellence is late arriving. For some reason, GM seems to be leaving a lot of its development work to its first round of customers. This phenomenon killed the Pontiac Fiero and has caused Allanté sales to fall below expectations. That's a shame because the current Allanté is both a good car in its own right and a considerable value compared with the increasingly pricey Mercedes-Benz 500SL. —*Jack R. Nerad*

CADILLAC ALLANTÉ

Is the Allanté finally an enthusiast car?

PHOTOS BY LESLIE L. BIRD

CADILLAC'S ALLANTÉ HAS never before made the cut when finalists for *Sports & GT Cars* were selected. Its omission wasn't exactly deliberate but, when the time came to weed out the real enthusiast cars from the pretenders, not many people seemed to really like the car, much less have any passionate attachment to it. As a consequence, the Allanté became a perpetual Face on the Cutting Room Floor around here. But constants and absolutes are rare, and times—not to mention people and ideas—change.

The Allanté has changed. Though

still a *boulevardier's* dream of plushiness, size and exclusivity, the Cad has metamorphosed into something of a real GT car during the last two seasons. For 1990, thanks to some additional wiring and a pound or two of extra hardware, the Allanté is finally ready to play among the world's best Grand Tourers.

Not a minute too soon, either. General Motors has not been known for an eagerness to sustain low-volume cars when they don't add to corporate profits. The Cadillac sporty car had to be a concern for GM bean-counters (Chevrolet probably loses more cars in a year than come down the Allanté as-

sembly line). That may change.

Let me be clear on one vital point: the Allanté was never a *bad* car. It always has been comfortable, well-built and easy to look at; just not a whole lot of fun to drive, if your idea of motoring pleasure places more demand on a car than a run to the golf course every other Saturday.

Our automotive thrills, however, extend a bit beyond loping out to the links, and that's where the Allanté has heretofore fallen short. Taken over a challenging piece of pavement, previous Allantés have been strictly among the also-rans.

Again: The Allanté has changed. A 1990 Allanté, thanks to suspension revisions, needn't be humbled by little buzz-bombs at play. Cadillac engineers have refined shock absorbers, anti-roll bars and other chassis parts to a fine edge, to the point where the 1990 model steers, grips and handles like a "real" sports car, without any sacrifice of ride quality. Torque steer, a past problem, is virtually nonexistent. The brakes were excellent before, and still are. And the Allanté will accelerate at an immodest pace when necessary. Sound like great fun? It is. Certain Germanic rivals may keep up or, in a few cases, be a hair faster, but it'll be a real contest.

That's the normal scenario; the situation will change dramatically in Al-

> **You can't help but approve of the alloy V-8 engine and its blend of silkiness at cruise and bellowing full-throttle response . . .**

lanté's favor during hard driving when the road is wet, or when there is no pavement, as on gravel or dirt.

The secret is Traction Control, a mechanism that Cadillac can be justly proud of adopting. Read the accompanying article for the *how* of its operation; let's talk about results. Assume the application of too much throttle in a tight turn. Where wheelspin would be the ordinary result, an Allanté driver will experience the uncanny sensation of full acceleration, if at a fractionally slower rate, without having to feather the throttle or be concerned with unwanted steering wheel movement. Speed within the corner may not be higher, but the exit velocity surely will be.

Same goes for runs in the wet or other slippery conditions. Rather than causing loss of stability, full power brings the system into play, which then doles out only that amount of torque that the tires can transmit groundward. You can learn a new driving style in this car, one that's certainly easier on the driver and, if you are among that group of us who aren't as fast as we'd like, you'll gain some speed in the process.

Traction Control has an even more important function: *safety*. This is the biggest advance since ABS in the war against unpleasant surprises on the road, to my mind, and I hope it spreads to other cars quickly. Even in normal driving, the world is full of situations where loss of grip can have disastrous consequences. A 200-bhp front-wheel drive machine wouldn't ordinarily be what I'd recommend you send your loved ones into a rainy night in; this is the sole exception. Hard as I tried, I could find no fault with Traction Control; it's unobtrusive in operation, effective, and sure to be reliable as the sunrise.

But! Like ABS, this system does *not* perform miracles. Don't jump to the conclusion that it somehow produces traction where there can be none, or will save you from blatant stupidity. What Traction Control *does* do is take worry about wheelspin out of the driving equation, enabling a skilled pilot to press harder, or an ordinary driver to avoid some mishaps. It does not replace attention, skill and a measure of caution in unfamiliar situations.

End of sermon.

Once we're past the better suspension and Traction Control, there's little new to talk about, though what's been done is all for the better. New seats are easier to adjust and more relaxing to sit in, a revised steering wheel holds an airbag and a CD player is added to the sound system.

Those parts that continue unchanged still deserve kudos. You can't help but approve of the alloy V-8 engine and its blend of silkiness at cruise and bellowing full-throttle response, or the crisp-shifting 4-speed automatic transmission attached to it. Then there's the lovely exterior—a Pininfarina product—simple, elegant, tastefully accented with bright trim and superbly finished. Though larger than a 2-seater need be, the Allanté's alluring good looks come as close as any modern design for timeless appeal.

Elegant design fills the Allanté's interior. Six large and graphically clean dials fill the instrument panel, large pushbuttons control virtually every driver-ordered function, and the climate control system is typical GM fare, being very effective. Leather, top-quality plastic and fine carpeting, carefully assembled, give the cabin an air of opulence. It's an easy environment to get used to.

However, there are a couple of nits to be picked. For starters, two gauges—oil pressure and water temperature—are obscured by the new steering wheel. Considering the size of the instrument panel, that's unnecessary. But the main offender is the top: Front latches are awkwardly placed, the motorized attachment below the rear window that holds the top in place is hit-and-miss in operation and, worst of all, the whole unit, when lowered, is covered by an unsightly and hard-to-close plastic panel. Others have created simple-to-deploy tops; it should not be beyond Cadillac to do the same.

Just for (derisive) laughs, there are the gold-colored ignition and door keys, or the multifunction trip computer that spells out "Good Morning" (or Afternoon or Evening, as the case may be) when the engine's started. Or the key-in-ignition *cum* seatbelt warning, a perfect department store elevator bell ("Third floor, ladies' lingerie")

soundalike. Cadillac marketing seems to revel in these touches.

But this is a Cadillac like no other. Traction Control is one factor (albeit a major one) in its appeal, but other praiseworthy elements are present too. The Allanté is short on flash and long on tasteful appointments, pleasing to look at. Above all, it's a delight to drive, in town and out beyond nowhere, on those roads you take for fun instead of shortest-distance travel. Call it a true dual-purpose Grand Touring car that is comfortable, refined, full of sound and fury but without wheelspin.—*Ray Thursby*

PRICE

List price, all POE $50,900	Price as tested $50,900

Price as tested includes std equip. (air cond, elect. window lifts, elect. adj mirrors, trip computer, elect. adj seats, ABS)

ENGINE

Type	ohv V-8
Displacement	4467 cc
Bore x stroke	92.0 x 84.0 mm
Compression ratio	9.0:1
Horsepower, (SAE)	200 bhp @ 4400 rpm
Torque	270 lb-ft @ 3200 rpm
Maximum engine speed	5200 rpm
Fuel injection	electronic port
Fuel	unleaded, 92 pump oct

GENERAL DATA

Curb weight	est 3465 lb
Test weight	est 3615 lb
Weight dist, f/r, %	est 56/44
Wheelbase	99.4 in.
Track, f/r	60.4 in./60.4 in.
Length	178.7 in.
Width	73.5 in.
Height	52.2 in.
Trunk space	16.3 cu ft

DRIVETRAIN

Transmission .... 4-sp automatic

Gear	Ratio	Overall ratio	(Rpm) Mph
1st	2.92:1	9.37:1	na
2nd	1.57:1	5.04:1	na
3rd	1.00:1	3.21:1	na
4th	0.70:1	2.25:1	est 130

Final drive ratio 3.21:1

CHASSIS & BODY

Layout	front engine/front drive
Body/frame	unit steel
Brakes, f/r	10.3-in. vented discs/10.0-in. vented discs, vacuum assist, ABS
Wheels	forged alloy, 16 x 7
Tires	Goodyear Eagle VR, 225/55VR-16
Steering	rack & pinion, power assist
Turns, lock to lock	2.4
Suspension, f/r	MacPherson struts, lower A-arms, coil springs, tube shocks, anti-roll bar/Chapman struts, lower A-arms, transverse fiberglass leaf spring, tube shocks

ACCELERATION

Time to speed	Seconds
0–30 mph	2.7
0–40 mph	4.0
0–50 mph	6.2
0–60 mph	8.3
0–70 mph	11.3
0–80 mph	15.3
0–90 mph	20.5

Time to distance	
0–100 ft	3.3
0–500 ft	8.8
0–1320 ft (¼ mi)	16.4 @ 82.8 mph

BRAKING

Minimum stopping distance	
From 60 mph	142 ft
From 80 mph	259 ft
Control	excellent
Overall brake rating	very good

FUEL ECONOMY

Normal driving	na
EPA city/highway	15/22 mpg
Fuel capacity	22.0 gal.

HANDLING

Lateral accel (200-ft skidpad)	0.82g
Speed thru 700-ft slalom	61.7 mph

Subjective ratings consist of excellent, very good, good, average, poor.
na means information is not available.

Cadillac Allanté

Twelve months: What do you get when you combine Italian styling with American engineering? Mixed feelings, mainly.

FOUR SEASONS TEST

Ann Arbor—

When our bright red Cadillac Allanté was still new to our Four Seasons fleet, Michael Jordan asked the following questions in his initial report: "What do you really get for nearly $60,000? Is this a real American car or a marketing gambit? What will happen when Mercedes introduces its new SL and BMW announces its new 8-series coupe?" As the miles piled on, some of the questions were asked again by the people who wrote entries in the logbook. After a year and more than 23,000 miles, it's clear that some of the questions will have to go unanswered.

For one thing, where *does* the $57,183 price fit in with Cadillac's image? Is it a premium paid for the expensive "air bridge" routine (Allanté shells are air-freighted from Pininfarina's assembly facility outside Turin to GM's Hamtramck plant for powertrain installation), or does the division really see the car as the equivalent of upscale European roadsters? Now, a year later, the car can be spared comparisons with new Mercedes and BMW two-seaters. Their prices have gone so ballistic that they're entirely in another orbit. In fact, they flatter the Cadillac's sticker by contrast.

It would be terribly easy to take a leaf from the largely unimpressive sales record book and flay the Allanté with it, but the fervent conjecture that fills the comments section of our logbook tells a better story. In it are exuberant tales of cross-country adventure, high-speed tests, and boulevard posing as well as some of the bitter frustration at electronic treachery and roof mechanism intractability. Even as we note this, we are aware that Cadillac's own divisional dossier documents myriad development upgrades. The specter of GM's reputation for letting its customers do the final testing work really comes back to haunt the company on this car.

But that doesn't mean our 1989 Allanté was a car that could not be enjoyed. On the contrary. The 200-bhp V-8 that was tweaked anew for this model year by GM engine guru Jim Queen (and wrongly credited to Jack Roush in this and other publications) has continued to charm drivers with its unflagging enthusiasm, even if the decided lack of throttle progressivity has not. From the outset, this car was troubled by an accelerator pedal that would yield smoothly to the pressure of a shoe for about an inch and a half and then hang up against some unknown resistance. The increased pressure needed to overcome the blockage would invariably overshoot the throttle deep into the footwell, whereupon Jim Queen's responsive 4.5-liter V-8 would suddenly peel its tires.

This could be embarrassing, especially in front of people and police officers on whom you were trying to impress your sense of responsibility and poise. Experiences in later, smoother Allantés make us now wonder if we might not have had the linkage adjusted, freed, or fixed in the interest of more discreet departures. An easing of the throttle linkage probably would have reduced the heavy front tire wear, too, which resulted in replacement by nice new all-season Michelin XGT V4 covers at a point not much beyond 20,000 miles. However, the timing was good, as we were by then in the grip of the Michigan winter,

BY BARRY WINFIELD

PHOTOGRAPHY BY ROB JOHNS

CADILLAC ALLANTÉ

and the improved performance of the aggressively patterned XGT V4s provided the Allanté with enough traction to tackle snow-covered rural roads.

Recent models also render our oft-voiced criticisms of the seat adjustment buttons somewhat irrelevant, as the people at Cadillac have made this year's Braille much easier to figure out. Actually, if this car were someone's everyday transport, even the vagaries of the '89 switch layout would soon become second nature. Our drivers were certainly fairly tolerant of the console push buttons, which—despite their unvarying color and shape—were the right size for a gloved finger. The disparity in lighting levels between the analog instruments and the yellow-backlit information panel and radio controls troubled most of us, usually resulting in the instrument lighting's being dimmed to a scarcely visible gloom to reduce the yellow glare.

Despite this car's automatic three-position damping control, the ride was routinely castigated for being harsh, abrupt, and violent on rough surfaces. Its poise and balance on smoother pavement went largely unremarked, but it should be recognized that the car has a suspension setup much more like a Corvette's than a Fleetwood's and is better confined to the smooth asphalt of the Sun Belt than the cratered thoroughfares of the north. With body motions held pretty well in check, and with tires of a distinctly generous cross-section, the Allanté has secure roadholding and prodigious grip. Unfortunately, the variable-assist power steering is still too numb to deliver messages of sufficient quantity or fidelity to inspire real cornering confidence, and the frequent directional inputs resulting from surface imperfections, camber fluctuations, and plain old torque steer also compromise the driver's sense of security.

If the styling is generally admired (and it is) and the performance not found wanting, then the operation of the fabric top was in sharp contrast. Drivers complained that even following the directions step by step was not enough to guarantee success. On the other hand, at

REACTIONS

An attractive proposition, but still in need of refinement.

Cadillac has chosen a very expensive way to build the Allanté. Components are flown to Pininfarina in Turin aboard special Boeing 747s, where bodies are built and then shipped back for final assembly in Michigan. It is quite possible that the Allanté will never turn a profit, even at its high price, simply because the manufacturing process is so complex and inefficient. Pininfarina is justly famous for creating bodies that are beautiful—and the Allanté's body is certainly a thing of beauty—but quality is another matter, and neither Pininfarina nor the '89 Allanté is apt to win any Pulitzer Prizes for fit and finish.

Our Four Seasons car was beset by creaks and rattles before we'd had it for six months. Electric thingies died. The top was difficult to raise and lower. The throttle was sort of an on-or-off proposition. Average owners probably did better than we did, because they'd be less apt to do almost 25,000 miles in the first year, but even so, there were too many bugs and annoyances for a car in this class.

I recently drove a 1990 Allanté. Like many GM products in recent years, the car improves steadily as the model years roll by. The '90 model was much more refined than our test car and featured a broad range of improvements, not the least of which was traction control—noisy, but effective. If the whole car were built in Hamtramck (and the resulting savings were passed along to the customer) and the refinements kept coming, it wouldn't be long before the Allanté would be a very attractive proposition.
—David E. Davis, Jr.

I'm used to driving small cars. So on my initial meeting with the Allanté, my overwhelming impression was of its size—nearly fifteen feet long and 3500 pounds. Sure, the Allanté is a two-seater, but it's also a Cadillac. And in the Cadillac tradition, it's not a small car.

Not that there's anything wrong with a car being large. After all, this is a luxury car, and for nearly $60,000 people expect a lot. With the Allanté, they'll have attention-getting exterior styling and a powerful V-8 engine. And along with a big car comes a good-sized trunk, which makes for a great weekend tourer.

But in the Allanté, even after fiddling with the myriad seat controls, I couldn't get the driving position right. Maybe it had something to do with the low-slung, hard seats. Or maybe it was the fact that the rear window's sill was so high that all I could see when I turned my head to look out the back was the inside of the car. This last problem wasn't helped by the operation of the side-view mirrors. Because the mirror motors had such a limited range of movement, it often became necessary for drivers to reach out the windows and move the mirrors manually before making fine adjustments with the power switch. At least one staffer never figured this out, always driving with the mirrors reflecting the door handles or the sky.

Cadillac could give the Allanté softer seats, greater all-around visibility from the driver's seat, and fully adjustable remote mirror controls, and then I'd not only look good while driving the Allanté, I'd feel good, too. —Amy Skogstrom

Back in the mid-Fifties, when the well-off middle-aged women who are the target audience for the Allanté were dewy-cheeked young girls in poodle skirts and bobby socks, a Cadillac convertible was a wonderful piece of technology. If you left it parked outside your club with the top down, you didn't need to worry about a sudden shower; raindrops on a sensor plate between the front seats would activate a completely automatic sequence that put the top up and rolled up the windows. Why, then, did Cadillac management ever think that Americans could accept—should even *tolerate*—the miserable excuse for a top foisted on them by Pininfarina?

Crude and obstructive from the start, still hopeless after annual "improvements," the Allanté's folding top is a joke. I suspect it has done more to keep sales below anticipated levels than any of the car's other shortcomings, most of which have been nicely overcome by development work that should have been completed before the first car was sold.

The styling may be conservative, but it *does* attract attention—favorable attention, I would stress. Curiously, the Allanté was not often recognized as a Cadillac; perhaps the squared egg-crate grille is not on its own enough of an identity factor. Weighed against the eighteen-year-old Mercedes SL, the Allanté is a reasonable "speedy roadster." Compared with the current SL, it is woefully substandard, and its lower price doesn't help. Cadillac should simply design a new Allanté all by itself. One with a power top. —Robert Cumberford

CADILLAC ALLANTÉ

least two people noted in triumphant tones that a little familiarity bred a whole lot of technique, making the job a snap. Be that as it may, the car required a new power pull-down latch at 13,000 miles. Perhaps our last words on the subject should be that the 1990½ model wears yet another, revised version of the roof, and that our experience of it during our eight-car convertible test [May 1990] did not win many new admirers.

As for general fit and finish and mechanical dependability, we have to score those somewhat short of perfection. The fabric top has frequently leaked, and even the hard top we fitted for winter emitted drips and whistles from the window tops. Early electronic problems (refusing to start, persistent warning of low oil levels) were taken care of under warranty—although it took two tries to solve the false low-oil-pressure alert—and have not returned in any form to plague us of late. The general appearance of the interior divided opinion about as far as it can go, with one writer saying that everything looked and felt expensive, and another saying the interior required a complete demolition and rebuild job. Similarly, the hard seats were accused of being responsible for day-after aches and pains by one passenger and considered comfortable enough to be cut out and used as living-room furniture by another.

The large and thin-rimmed steering wheel drew some negative responses, and the stubbornly recalcitrant glove box and console cover mechanisms—which require brutal violence to latch—seemed totally out of character for a prestige car. The mirrors and their adjustment mechanisms (manual and power) have also been the targets of fairly frequent criticism. Plaintive whines about a stiff gear selector can only be entertained in light of the Allanté's likely fastidious target market. Another niggle is that the central locking mechanism tends to close on your fingers like a mousetrap if you're not careful.

For all that, the Allanté is a likable, stylish car, comfortable and fast on long trips, and it always drew plenty of attention. A pity that the attention it needed—to the details that could have made it a truly classy car—was unavailable. As automotive design editor Robert Cumberford aptly observed: "The car would be a better deal at $30,000, but I would *still* jump on them for the top, the mirrors, and the concrete seats."

CADILLAC ALLANTÉ

Despite a stiffish ride with quite a bit of road noise on poor surfaces, the Cadillac Allanté's roomy cabin and respectable luggage space made it a popular long-distance tourer. It was during these cross-country jaunts that its exotic bodywork and little-known identity came into focus. Most people did not know what it was, but they seemed to like it. Most of us liked it, too, or at least parts of it. The hearty V-8, the flamboyant red coachwork, and the luxurious appointments were hard to fault. Conversely, recurrent electronic glitches, mirror motor failures, and difficulties with the Rube Goldberg soft-top mechanism tried everyone's patience, even though these problems were dealt with under warranty. We paid only for tire valves, balancing, lubrication, filters, and quite a bit of fuel. At the end of its stay, the Allanté was looking good and running as well as ever, and we'd like to think it was set for a long and trouble-free future. At the time of its acquisition, the car's sticker asked $57,183. It's now likely to fetch less than $45,000.

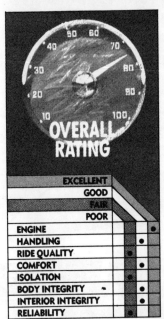

REVIEW PERIOD:
23,208 miles
Previous articles October 1989, May 1990

SCHEDULED MAINTENANCE:
Parts $19.79
Labor $24.00
Fluids $30.53

WARRANTY REPAIRS:
Parts $597.56
Labor $398.00 (calculated at $44 per hour)

NONWARRANTY REPAIRS:
Parts $50.04
Labor $128.00

FUEL ECONOMY:
EPA city 15.0 mpg
Our observed 18.3 mpg

PROBLEM AREAS:
Top operation
Oil level sensor
Oil leaks
Window seals

CADILLAC
ALLANTÉ

Wish upon a rising star

PHOTOS BY DEAN SIRACUSA

For many executives in Detroit, this year's holiday festivities will be tinged with anxious anticipation. The issue that has the engineers and product planners of Clark Street drumming their fingers on desktops is the long-rumored, long-awaited Northstar engine, an all-new 32-valve V-8 that will grant Cadillac's newest products, the Seville and Eldorado in particular, that final element of mechanical pedigree they now deserve.

But before this diadem of the new Detroit enters the mainstream, it will undergo its first public scrutiny in the Allanté, the Italian-American hybrid that Cadillac and Pininfarina introduced five years ago to spar with the Mercedes SL. Of course, everyone wants to know just how good the Northstar really is, and all we can offer for now is a preliminary prototype evaluation (see p.36). Yet one thing is certain: Cadillac is quite serious about implementing new technology, and feels the Allanté is the place to test it.

That priority has won the Allanté a steady diet of firsts for GM during its life, including an electronically controlled transmission, automatic ride control, variable-boost power steering and, most recently, traction control. There's no question that adding a 32-valve V-8 to this list will contribute to the Allanté's attractiveness, but the current car remains appealing in its own right. First and foremost is its styling, which, though somewhat dated in comparison to the 300 SL and 500 SL, possesses an element of timelessness that should maintain a strong appeal years hence. There is much to like here in the strong character lines that meet at softly burnished junctions, in the strong "shoulders" that build from a delicately tapered nose. Moreover, Turin's tradition of metalcrafting has resulted in smooth, precisely joined panels covered with a glass-like finish that rivals any luxury car from Japan or Europe. Best of all, Cadillac has greatly limited the jewelry the Allanté must wear, so the overall look is pleasantly understated.

Pininfarina tradition also shapes the Allanté's interior, but the lines there may be less likely to achieve classic status. At a time when organic shapes have sprouted inside lower-priced family sedans such as the Buick Skylark and Ford Taurus, the Allanté's angular design seems out of step with the Nineties. But don't be too quick to judge it as retrograde. Unless you choose the gee-whiz electronic instrumentation, the standard analog gauges with old-fashioned numbers and long, thin needles are perfectly fine. And with its power output raised to 200 watts, the Delco/Bose Gold Series audio system with standard CD player can now impress not only the neighbors, but also the entire neighborhood with the finest Mozart—or Metallica. Less impressive is the profusion of square and rectangular buttons used to regulate all that sound, as well as the climate control system and trip computer. Everything generally makes sense here, but finding the appropriate button without taking your eyes off the road poses a challenge. Let's hope an upcoming facelift will simplify the design.

What Pininfarina shouldn't change are the lovely Recaro bucket seats whose supple "Durosoft" leather hides fill the cockpit with the expensive aroma of an i Santi briefcase. With Cadillac resisting the urge to offer yet more buttons, the Recaros can suit any size driver with a wide range of adjustments using traditional seat-side controls and an assist from the adjustable steering column. Contributing to the feel of self-indulgent luxury is what may be the richest carpeting in any American car—it's enough to make you want to drive barefoot.

Since the Allanté's introduction, Cadillac's gone to extreme lengths to add power assistance to its top without actually offering a true power top. While the first-generation design received almost universal disdain—especially considering that its creator designed classic one-handed tops for Alfa and Fiat—the current Phase II approach seems almost foolproof. Although the top stack must still be raised and lowered manually, the front and rear headers are anchored electrically, so the whole operation now takes less than a minute to perform. But with the Mercedes SL and the Infiniti M30 now offering automatic tops whose opera-

tion is as mindless as a microwave, the onus is on Cadillac to create a one-touch Phase III design as soon as it can get one into production. Before that takes place, however, Allanté buyers will just have to fold for themselves.

Marking time in the engine compartment until the introduction of the Northstar engine is the current overhead-valve V-8. Although its displacement remains at 4.5 liters when other Cadillacs have jumped to 4.9, its tuned-port induction system matches the 4.9's 200-bhp output, getting the Allanté's 3600 pounds to 60 mph in a scant 8 seconds. While that's a second and a half slower than the 2-ton 500 SL, it's certainly fast enough—while gaining a few mpg in the process. With an electronic brain calling the shots, the 4-speed auto shifts smoothly, even under full-throttle starts. Torque steer remains fairly mild in such conditions, limited no doubt by the Allanté's traction control system, which uses a combination of braking and engine management to limit wheel slippage.

Like the SL, the Allanté's not a sports car, although Cadillac has gone to great lengths to prevent it from being perceived as a *boulevardier*. The automatic ride control and variable-boost steering firm things up with speed. Not that the Allanté's manners are ever sloppy: Most drivers will find its ride and handling far stiffer than expected, particularly once speeds increase. The transition from one set of control conditions to the other is subtle, but the bottom line is that on the highway, the Allanté has a strong on-center feel that keeps it tracking straight. The reduction in steering boost also enhances high-speed cornering, which means that the Allanté isn't afraid to take that favorite mountain pass at a brisk pace. Assisting in the process are the 225/55-16 Goodyear Eagles, whose generous grip won't disappoint more ambitious drivers.

As the last performance of a play whose cast may be significantly changed when it reopens in 1992, the first-generation Allanté remains a highly attractive, albeit specialized, automobile. While it may lack some of the integration of power and performance that Mercedes' SL enjoys, it compensates with a uniquely Italian interpretation of elegance and luxury, executed with pride and class. There are some carmakers—notably Chrysler—who thought that simply building a car in Italy (Chrysler's TC by Maserati) and draping it with a half-dozen leather hides would buy them that class. But Cadillac knows otherwise. Getting the Allanté off the ground and steadily updating it from year to year has required patience, persistence and plenty of cash. The result remains a sophisticated vehicle, perfectly poised for a fresh injection of vitality.

—*Lowell C. Paddock*

CADILLAC ALLANTÉ

PRICE
List price, all POE .... $57,140   Price as tested $57,140
Price as tested includes std equip. (climate control, elect. window lifts, central locking, AM/FM stereo/cassette/CD player, elect. adj seats, elect. adj mirrors, traction control, ABS)

ENGINE
Type	ohv V-8
Displacement	4467 cc
Bore x stroke	92.0 mm x 84.0 mm
Compression ratio	9.0:1
Horsepower, (SAE)	200 bhp @ 4400 rpm
Torque	270 lb-ft @ 3200 rpm
Maximum engine speed	5200 rpm
Fuel injection	electronic port
Fuel requirement	unleaded, 92 pump oct

GENERAL
Curb weight	3552 lb
Test weight	3615 lb
Weight dist, f/r, %	61/39
Wheelbase	99.4 in.
Track, f/r	60.4 in./60.4 in.
Length	178.7 in.
Width	73.5 in.
Height	51.2 in.
Trunk space	16.3 cu ft

DRIVETRAIN
Transmission .... 4-sp automatic

Gear	Ratio	Overall ratio	(Rpm) Mph
1st	2.92:1	9.37:1	32
2nd	1.57:1	5.04:1	51
3rd	1.00:1	3.21:1	79
4th	0.70:1	2.25:1	(4000) est 130

Final drive ratio 3.21:1
Engine rpm @ 60 mph in 4th 1850

CHASSIS & BODY
Layout	front engine/front drive
Body/frame	unit steel
Brakes, f/r	10.3-in. vented discs/ 10.0-in. vented discs, vacuum assist, ABS
Wheels	forged alloy, 16 x 7
Tires	Goodyear Eagle VL, 225/55VR-16
Steering	rack & pinion, power assist
Turns, lock to lock	2.4
Suspension, f/r	MacPherson struts, lower A-arms, coil springs, tube shocks, anti-roll bar/ Chapman struts, lower A-arms, transverse fiberglass leaf spring, tube shocks

ACCELERATION
Time to speed	Seconds
0–30 mph	2.7
0–40 mph	4.0
0–50 mph	6.2
0–60 mph	8.3
0–70 mph	11.3
0–80 mph	15.3
0–90 mph	20.5

Time to distance	
0–100 ft	3.3
0–500 ft	8.8
0–1320 ft (¼ mi)	16.4 sec @ 82.8 mph

BRAKING
Minimum stopping distance
From 60 mph .... 142 ft
From 80 mph .... 259 ft
Control .... excellent
Overall brake rating .... very good

FUEL ECONOMY
Normal driving .... 16.6 mpg
EPA city/highway .... 15 mpg/22 mpg
Fuel capacity .... 22.0 gal.

HANDLING
Lateral accel (200-ft skidpad) 0.82g
Speed thru 700-ft slalom .... 61.7 mph

*Subjective ratings consist of excellent, very good, good, average, poor.
na means information is not available.*

1993 Allanté With Northstar V-8: Subtle Change, Big Improvement

Over the past half-decade, the Lotus-developed Corvette LT5 has been the sole American response to five new multivalve V-8s from Europe and Japan. But now the time has come for America to leave the technological sidelines, and it's only fitting that Cadillac, developer of the first high-compression V-8, should be first off the bench. Its new powerplant is named "Northstar," for its leading role within GM, and next spring it will offer front-drive Cadillacs, starting with the Allanté, a substantial improvement in performance and refinement.

The alluring Allanté has always been Cadillac's technical testbed, and with Northstar that means, among other improvements, a 4.6-liter alloy block fitted with iron liners, free-floating forged aluminum pistons, chain-driven cams with hydraulic tappets, a tuned thermoplastic intake manifold, computer-controlled port fuel injection and distributorless direct ignition. The result is 290 bhp at 5600 rpm, a whopping 90-bhp gain over the 4.5-liter ohv design. Torque benefits as well, though by a narrower margin: 290 lb-ft at 4400 rpm versus 270 at 3200. That makes the Allanté the most powerful fwd car in the U.S., a challenge it tames with a new transmission that blends electronic engine dialog and mechanical improvements to smooth shifts and reduce internal friction.

While Northstar helps the Allanté gain ground on the 322-bhp 500SL (Cadillac claims a 0-60 of 7.0 sec, vs.

6.4 for the 500SL in our testing), a new adjustable damping system swaps the previous all-wheels-at-once design for one that adjusts each wheel independently. Improving handling is a new independent rear suspension that mates coil springs and unequal-length upper and lower arms with a lateral link to control toe angle during suspension movement, reducing roll-induced oversteer. The new setup also minimizes brake dive on high-g stops. Finally, a new speed-sensitive steering system more smoothly proportions boost based on data from the same microprocessor that controls damping.

Given that the Allanté, from a dynamic perspective, is virtually a new car, Cadillac limited its interior and exterior changes, reasoning that the marketplace says it likes both as is, and the cost of the change couldn't be justified. "We don't run the Allanté like a charity," says Dave Hill, chief engineer for Allanté and Fleetwood. "People still point at the car, they like the way it looks. We don't need to beat them over the head with a new model."

That said, a myriad of detail changes have made the car more livable, such as the deletion of the vent panes in the door glass, allowing the mirror to be moved forward to a more natural position. Equally functional but less aesthetically pleasing is a new chin spoiler that reduces high-speed lift. Opening the doors is now easier, and inside redesigned seats receive a lumbar support that's adjustable both vertically and horizontally. There's even a pair of the now-ubiquitous cupholders, should you wish to take your Perrier along.

So does the whole exceed the sum of its parts? During a brief comparative stint with the 500SL I'd have to say yes. Apart from the silky thrust generated by its new engine, the most apparent improvement is the car's "rolling feel." Whereas the previous car had a comparatively firm ride and numb steering response, its replacement is more supple overall, offering greater road isolation, vastly improved ride quality, better directional stability and more responsive steering. The billet-like SL still excels in chassis stiffness, however, an advantage gained at a 400-lb premium over the Allanté.

Given that it will sell for, say, $25,000 less than the $90,000 SL, the 1993 Allanté is definitively more competitive than before. But will consumers look past its unchanged exterior and savor the technology below? That's a marketing hurdle something more substantial than Cadillac Style will have to surmount. —L.P.

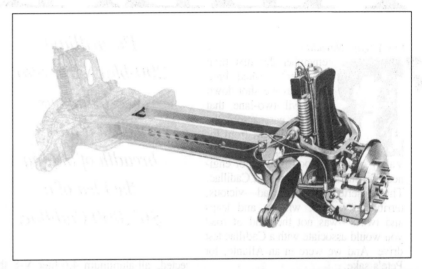

DRAWINGS BY DAVID KIMBLE

1993 CADILLAC ALLANTÉ

Las Vegas, Nevada—

"Remember the first time we met?" asked Fred Wood as we shot down the evil two-lane that bounds east across the Mojave Desert from Interstate 15 toward the Valley of Fire. Wood is a senior engineer in vehicle development at Cadillac. This deserted Nevada road—vicious, terrible pavement, whoops and leaps and twists—was not the kind of road you would associate with a Cadillac test drive. And we were in an Allanté, for Pete's sake.

"It was three years ago, and you asked me if we were really serious about this car. We're serious about this car, Jean," said Wood.

I had just done this nasty twenty-mile leg in a Mercedes-Benz 500SL and had bottomed its suspension in the very dip that now faced me and the 1993 Allanté. I clenched my teeth, hit that hole at 80 mph, and waited for the sickening crash-through. Wood was smiling serenely over in the passenger's seat because he was in charge of the Allanté's new rear suspension. We dipped. We squidged. We rebounded. No crash-through.

I thought I had flown out to Las Vegas to form a close personal relationship with the long-awaited North Star V-8, the engine that will eventually power the Seville and the Eldorado. I had come to meet Cadillac's first new engine in a decade, a lovely DOHC 32-valve, fuel-in-

The brilliant 290-bhp North Star V-8 brings us within a hair's breadth of buying the idea of a $67,000 Cadillac.

jected, all-aluminum 4.6-liter V-8 that promised nearly *50 percent* more power than the 200 horses of the 4.5-liter V-8 it was replacing.

Stop the presses here a second. It wasn't too long ago that engineers were excitedly talking about breaking the 200-bhp mark for a front-wheel-drive car. And what are we looking at? 290? And from what? A Cadillac!

Cadillac has been developing the North Star for five years. For the record, its sophisticated two-piece aluminum block is a design you find most often on dedicated racing engines, not production-car powerplants. It is composed of an upper cylinder block with cast-in iron liners, a lower ladder-frame assembly supporting the crankshaft, and an additional oil manifold plate that seals a groove in the lower crankcase to form the oil gallery.

Extensive work was done to suppress noise. A roller chain drives the four overhead cams (two per cylinder bank), the starter housing sits in the crook of the engine, and the computer-controlled multipoint injection system (Cadillac calls it the Fluid Induction System) is essentially a magnesium box that encloses the tuned runners, the fuel rail, and the injectors. This and other work resulted in an engine so quiet that a starter interlock was added to prevent the driver from inadvertently cranking the key while the engine is running.

So this is what I had come to town for: 290 horsepower and a lonesome desert highway blast to Salt Lake City as fast as I can fly it. (Cadillac hasn't completed testing but has seen less than 7.0 seconds from 0 to 60 mph and a top speed of 150 mph.) But no. This has turned out to be anything *but* a flat-out engine test. We weren't thirty-five miles north of Vegas, and we were off in the badlands, beating the hell out of a two-seat luxury convertible, number 00001 off the Allanté pilot line. Oh, the cockiness of these Cadillac engineers. Oh, the competence of their work.

As it turns out, the achievement of North Star dictated that almost the entire Allanté be overhauled: new transmission, new suspension, new brakes, new steering, new tires.

The 4T80-E transmission takes electronic shift control one step further. It can monitor the length and time of shifts and correct itself in the case of temperature change and wear on the compo-

....

BY JEAN LINDAMOOD

ILLUSTRATION BY DAVID KIMBLE

1993 CADILLAC ALLANTÉ

GENERAL:
Front-engine, front-wheel-drive convertible
2-passenger, 2-door steel and aluminum body
Base price (estimated) $67,000 (+ luxury tax of 10% over $30,000)

ENGINE:
32-valve DOHC V-8, aluminum block and heads
Bore x stroke 3.66 x 3.31 in (93.0 x 84.0 mm)
Displacement 279 cu in (4565 cc)
Compression ratio 10.3:1
Fuel system sequential multipoint injection
Power SAE net 290 bhp @ 5600 rpm
Torque SAE net 290 lb-ft @ 4400 rpm
Redline 6500 rpm

DRIVETRAIN:
4-speed automatic transmission
Gear ratios (I) 2.96 (II) 1.63 (III) 1.00 (IV) 0.68
Final-drive ratio 3.71:1
Traction control system

MEASUREMENTS:
Wheelbase 99.4 in
Track front/rear 60.4/60.4 in
Length x width x height 178.7 x 73.4 x 51.5 in
Curb weight 3766 lb
Weight distribution front/rear 63/37%
Ground clearance 5.7 in
Coefficient of drag 0.35
Fuel capacity 23.0 gal
Cargo capacity 16.3 cu ft

SUSPENSION:
Independent front, with electronically variable damper struts, coil springs, lower control arms, anti-roll bar
Independent rear, with electronically variable damper struts, coil springs, upper and lower control arms, anti-roll bar

STEERING:
Rack-and-pinion, variable-power-assisted
Turns lock to lock 2.7
Turning circle 40.2 ft

BRAKES:
Vented discs front and rear
Anti-lock system

WHEELS AND TIRES:
16 x 7.0-in cast aluminum wheels
225/60ZR-16 Goodyear Eagle GA tires

PERFORMANCE (manufacturer's data):
0–60 mph in 7.0 sec
Standing ¼-mile in 15.1 sec
Top speed 150 mph
Pounds per bhp 13.0

1993 CADILLAC ALLANTÉ

nents. The electronic torque converter is bigger, and the new aluminum transmission case has been reshaped to accommodate a larger final drive and differential. Equal-length driveshafts go a long way toward controlling torque steer, a crucial element in a front-wheel-drive car that is seeing 290 pounds-feet at its 4400-rpm torque peak.

The new rear suspension pieces—a short upper arm and a longer lower arm locating each wheel hub, together with an additional lateral link—are mounted on a subframe. The geometry of this multi-link system helps control oversteer tendencies and rear lift during braking and helps keep maximum tire patch on the ground during cornering and braking. Cadillac has also developed what it calls the Road Sensing Suspension, an electronic system that monitors vertical wheel velocity and body acceleration and can adjust shock absorber damping rates independently at each wheel in *real time*. At 60 miles per hour, the road is being read one inch at a time.

So the guys wanted to show off all this good stuff, in addition to humongous new brakes, a speed-sensitive steering system (controlled by the same computer that's running the high-tech suspension system), and new Z-rated 225/60ZR-16 Goodyear Eagle GA tires.

One thing led to another. The faster I charged toward the Valley of Fire, the harder I leaned on the suspension, and the better it worked. The harder I stood on the brakes—which were dynamite—the later I'd dive into the corners, reveling in the crisp steering, and the faster I'd go.

We returned to Interstate 15 eventually for that high-speed tear to Utah, and it was a gas. Power out of the hole, power at 50 mph, power at 80 mph, power at 9000 feet of elevation.

I found two flaws: At such a high altitude, the four-to-two downshift was a little harsh. (Karl Janovits, GM's engineering director of front-wheel-drive transmissions and one of the masterminds on the wonderful 4T60-E, called it "a hole in the calibration" and said it would be dealt with.) And I could fool the suspension if I hit a pothole with a sharp-edged drop. Ouch. That will also be scrutinized.

But here are two shining truths: The North Star engine is as smooth as a Lexus engine, with the performance excitement of an Infiniti's. And Cadillac is rare among the GM divisions in that it can build a high-performance car with a suspension that doesn't sink to its knees when you push it past 90 mph. The Allanté still doesn't have the structural rigidity of the 500SL, but Mercedes-Benz might want to look at the Allanté's suspension. And its price tag.

In fact, if I hadn't been staring at the cheesy-looking instrument panel and wall of control buttons that have plagued the Allanté's interior since its debut, I would have given the nod to its rumored-to-be-$67,000 price. (Some haggling is expected over this figure before the Allanté hits showrooms in the spring.)

Sure, there are other noteworthy improvements: The side-view mirrors are repositioned forward, a lower front spoiler was added to improve high-speed stability, and the window glass is now one piece that is 60 percent thicker for quieter and smoother operation. There are new seats (from the Eldorado), a redesigned top (no, not power, but new fabric), new wheels, and redesigned Mercedes-style seat controls.

But Cadillac's roadster is getting long in the tooth. One more round—in the design studio—and Cadillac will have it right. That will happen for the cockpit in 1994; a new skin is scheduled for 1996.

Cadillac *is* serious about this car.

CADILLAC ALLANTÉ
HIGH-ZOOT CRUISER

by Greg Coppock
PHOTOGRAPHY BY MIKE BANKS

LONG TERM ROAD TEST

Our '90 Cadillac Allanté has been like a long-term house guest you really enjoy having around, but who gets on your nerves from time to time. This high-profile roadster offers surprising performance, distinguished styling, and the creature comforts we've come to expect from Cadillac, but it also brings a slightly stiffer ride, button-happy ergonomics, and a frustrating convertible top.

The two-door, two-passenger Allanté is powered by a good, old V-8. The 4.5-liter mill thunders out 200 horses at 4400 rpm and has a peak torque rating of 270 foot-pounds at 3200 rpm; all of that is fed to the front wheels. Flex the right foot, hear the powerful growl, and you're at 60 mph in 7.9 seconds. As quickly as you got there, the ABS-equipped Cad brings you to a dead stop in a stable 125 feet. And with traction control, sticky 225/55VR16 Goodyear Eagles, and the deflected-disc adjustable shocks, the Allanté handles the skidpad with verve, pulling 0.85 g, and negotiates the slalom at 58.9 mph. The tradeoff for such wonderful performance is a stiffer ride than that of most Cadillacs. It's not harsh, mind you, just stiffer.

Cruising Beverly Hills in the Allanté was a kick all *MT* staff members enjoyed at one time or another. Anywhere the Cadillac cruised, it was a head-turner, but with its pricetag of $51,000, it should be.

And in view of the price, we became increasingly frustrated with the Allanté's top. The soft-top is almost entirely manual, and the mechanical piece that cinches the top down was overly temperamental, as we often spent more than 20 minutes just trying to get the top up. We took the Cadillac to a dealership a few times hoping to correct the problem, but the fix was always temporary. Cadillac has promised more efficient operation on future Allanté soft-tops.

The leather bucket seats offer an array of adjustments that ensure your comfort whether driving the Allanté enthusiastically or leisurely, and with the two memory settings, the transition is a breeze. Other pampering amenities include automatic headlights, climate control, trip computer, and Bose stereo with CD. The Cad has these goodies along with 44 buttons for their operation, so familiarization is necessary.

Having put 14,662 miles on our long-term Allanté, we discovered that servicing the high-line Cad was a snap, as the first couple oil changes were covered by the Gold Key warranty. When we had the tires rotated and new front brake pads installed, servicing was still a snap—though pricey at $357. Our total gasoline expenditure was high, as well—we spent $1356.40 for 929.87 gallons. Overall, the Allanté averaged about 12¢ a mile to operate, which admittedly is lofty, but such is the cost of luxury transport—particularly with our gang of leadfooted drivers.

Base price	$50,900
Price as tested	$51,650
Total mileage	15,318
Test mileage	14,662
Fuel consumed	929.87 gal
Fuel cost	$1356.40
Average mpg	15.8
Additional oil/cost	2 qt/$6
Routine maintenance	No charge/Gold Key
Additional maintenance	$357.57/Rotate tires; front brakes; washer fluid; oil change
Problem areas	Convertible top not latching
Total operating cost	$1719.97
Operating cost per mile	12¢

1993 Cadillac ALLANTÉ

The V-8 Northstar shines & shines

BY KEN ZINO
PHOTOS BY JOHN LAMM

FINANCIAL CHAOS, EXECUTIVE adjustments, worker redundancies and plant closings—these tumultuous times at General Motors blur the focus. At the root of the chaos is "product," good, bad or indifferent cars, trucks, vans. The Cadillac Allanté represents the best that beleaguered GM engineering brings to the current automotive fray, so it's an important event.

Moreover, Allanté's hardware and software applications affect more than just one model. The V-8 engine appears in limited-production 1993 Eldorado and Seville touring cars. The independent rear suspension eventually follows the engine onto other platforms. Its traction control and electronic damping will grace sundry GM platforms as model years wax. So Allanté for 1993 takes on greater significance than its diminutive sales prospects. The most substantive change is Northstar, Cadillac's first use of a multicamshaft, V-8 engine. Engine output—at more than 63 horsepower and lb.-ft. of torque per liter—is majestic. Redline is a lofty 6600 rpm. No wonder Allanté is the quickest (6.7 seconds to 60 mph) and fastest (145 mph) production Cadillac we've tested, and that's in spite of its ample 3740-lb. curb weight. (Perhaps not overweight, though, compared with the Mercedes-Benz 500SL at 400 lb. heavier.)

All the requisite technical tidbits are present: Northstar is an all-aluminum, die-cast V-8 with cast-iron cylinder liners and double-overhead camshafts for each cylinder bank. The camshafts are there to activate four valves per cylinder via direct-acting hydraulic followers. (All that's missing from this heady technical brew is variable valve timing.) Given the complexity and competence of this 90-degree V-8, its presentation is a contradiction: Care has been taken to make the mechanical aspects of Cadillac's first new engine in a decade as unobtrusive as the complex internal code of its computerized controls. Both are as undetectable to the driver as quarks to a non-physicist. Vibration? None evident. Engine noise? Shielded and damped to oblivion, with only a trace of tenor saxophone evident occasionally from the stainless-steel exhaust system.

Northstar is extraordinarily quiet. The chain case covers are composed of layers of steel with a rubber sound insulator sandwiched between. Other anti-noise techniques abound: The eight electronic fuel injectors and thermoplastic intake runners are located in a sound chamber with a magnesium cover. The lower crankcase assembly uses 4-bolt main bearings that surround the crankshaft and bolt to the upper block. Exhaust pipes are dual-walled with a ceramic filler providing heat and sound insulation. The resulting silence requires a starter motor interlock that prevents engagement while the engine is running.

Quiet but decidedly quick, with strong performance readily underfoot: Part-throttle acceleration is as punchy as the torque output suggests. While peak torque occurs at 4400 rpm, a robust 246 lb.-ft. (85 percent) is available at only 2000 rpm. Barely a dollop of accelerator travel is needed to move things up tempo, hastening lane changes or merges with traffic. Unrestrained throttle application gets you to 60 mph in less than 7 sec. This—only a tick more than a Nissan 300ZX Turbo or 500SL—is surely quick enough to bring smiles to countenances of fans of pre-emissions American V-8 muscle.

After a day of running hard through desert and mountains, the trip computer registered an average of more than 17 mpg; extraordinary in this league and within my experience. EPA ratings of 14 mpg city, 21 highway work out to a $1700 energy-use (also known as "gas-guzzler") tax added to the $61,675 suggested retail price. Add in the 10-percent luxury tax for the transaction amount above $30,000, plus sales taxes and licensing fees, and Allanté's disbursement also could be described as muscular.

There is more finesse, however, than bulging muscle in the luxury package, particularly in transferring the newfound power to the pavement. Complicating matters, of course, is the front-wheel-drive configuration—the world's most powerful front-drive setup. A 4-speed, electronically controlled transaxle is up to the task. As engine output builds with an upshift pending, ignition timing is retarded until the gearchange is effected. Part-throttle shifts are insignificant events, while full-power ones are crisply executed.

The 128-kilobyte memory of the engine and transaxle control computer is adaptive, meaning that an aggressive driving style is complemented with higher-rpm upshifts. The maximum upshift point is 6600 rpm. If you pull the console-mounted shift lever (the release button under the T-handle feels a bit tacky) into one of the lower gears and nail the throttle, the computer shuts off fuel at 6700 rpm.

Helping to keep things pleasant here is the traction-control/anti-lock braking system, governed by yet another computer. Engineers have left some full-throttle wheel chirp in the calibration (Mercedes, please note), giving the driver a hint of the raw power available and providing an excellent launch for our timed acceleration runs.

This tire squeal of delight doesn't continue endlessly toward the terror section of the sliding scale. If the Goodyear all-season radials fail to hook up, the mind of the machine overrules the driver's more passionate soul. By monitoring front-wheel speeds, the computer (a Bosch unit) can switch from "go" to "whoa" mode with a combination of reduced power and braking.

Simultaneously—in less than 100 milliseconds, actually—the powertrain computer cuts the fuel supply to as many as five cylinders, depending on the wheelspin. Brake application of one or both front discs lags slightly, so engine torque is reduced before the binders are applied. Like that of the engine, this Bosch system's complex operation is invisible to the driver. It has another advantage, as it works at all driving speeds when wheelspin occurs.

An obtuse driver, slow to realize the peril of a particular throttle setting and road condition, has more help available from the 4-channel, dual-diagonal, 4-wheel disc brakes. With the higher engine output come increases in brake caliper and rotor size. Pedal feel is firm, race-carlike, with modulation easily accomplished by increasing foot pressure. Panic time? Stand on it fast and hard while you steer for a clear path.

Steering is much better than you might expect, especially with this much power available through the front wheels. Pushed to the limit in our slalom, the Allanté is the equal of the rear-

■ Northstar V-8 engineers seem to have spent as much time on acoustics as on the production of power; at 290 bhp, this engine has plenty of the latter. And the antics of torque steer have been virtually eliminated, no mean feat considering the sheer twisting forces involved.

1993 Cadillac ALLANTE

PRICE
List price, FOB factory **$61,675**
Price as tested **$65,723**
Price as tested includes std equip. (ABS, driver airbag, AM/FM stereo/cassette & CD player, air cond, leather interior, cruise control, traction control; pwr seats, windows & mirrors), pearl coat paint ($700), Calif. emissions ($100), luxury tax ($3248), gas-guzzler tax (included in base price) $1700.

MANUFACTURER
Cadillac Motor Car
2860 Clark Ave.
Detroit, Mich. 48232

TEST CONDITIONS
Temperature .... 75° F
Wind .... calm
Humidity 40%
Elevation .... 990 ft

0–60 mph .... 6.7 sec
0–¼ mi .... 15.0 sec
Top speed .... est 145 mph
Skidpad 0.81g
Slalom .... 59.7 mph
Brake rating .... very good

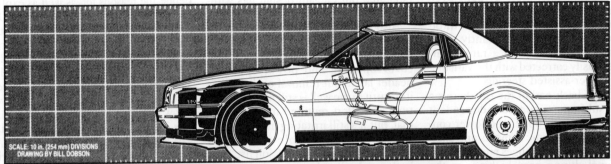
SCALE: 10 in. (254 mm) DIVISIONS
DRAWING BY BILL DOBSON

ENGINE
Type .... aluminum head & block, iron liners, **V-8**
Valvetrain .... dohc 4-valve/cyl
Displacement .... 279 cu in./4565 cc
Bore x stroke .... 3.66 x 3.31 in./
93.0 x 84.0 mm
Compression ratio 10.3:1
Horsepower
(SAE) ..... **290 bhp @ 5600 rpm**
Bhp/liter 63.5
Torque .. **290 lb-ft @ 4400 rpm**
Maximum engine speed .... 6600 rpm
Fuel injection .... GM elect. port
Fuel ...... prem unleaded, 91 pump oct

CHASSIS & BODY
Layout .... **front engine/front drive**
Body/frame .... unit steel
Brakes
Front ...... **10.9-in. vented discs**
Rear .... **11.1-in vented discs**
Assist type .... vacuum, ABS
Total swept area .... 375 sq in.
Swept area/ton .... 192 sq in.
Wheels .... alloy, **16 x 7**
Tires .... Goodyear Eagle GA,
P225/60ZR-16
Steering .... **rack & pinion,**
speed-sens pwr asst
Overall ratio 15.6:1
Turns, lock to lock 2.7
Turning circle .... 40.2 ft
Suspension
Front .... **MacPherson struts, lower A-arms,** coil springs, tube shocks, anti-roll bar
Rear .... **upper & lower A-arms, toe links,** coil springs, tube shocks, anti-roll bar

DRIVETRAIN
Transmission .... 4-sp automatic
Gear	Ratio	Overall ratio	(Rpm) Mph
1st	2.96:1	10.98:1	(6600) 45
2nd	1.63:1	6.05:1	(6600) 83
3rd	1.00:1	3.71:1	(6600) 138
4th	0.68:1	2.52:1	est (4710) 145

Final drive ratio 3.71:1
Engine rpm @ 60 mph in 4th 1950

GENERAL DATA
Curb weight .... **3740 lb**
Test weight .... 3900 lb
Weight dist (with
driver), f/r, % 62/38
Wheelbase .... 99.4 in.
Track, f/r .... 60.4 in./60.4 in.
Length .... 178.7 in.
Width .... **73.4 in.**
Height .... **51.5 in.**
Ground clearance .... 5.7 in.
Trunk space .... 14.1 + 4.0 cu ft

ACCOMMODATIONS
Seating capacity 2
Head room .... 39.0 in.
Seat width .... 2 x 21.0 in.
Front-seat leg room .... 44.0 in.
Seatback adjustment .... 35 deg
Seat travel .... 7.5 in.

INTERIOR NOISE
Idle in neutral .... 47 dBA
Maximum in 1st gear .... 77 dBA
Constant 50 mph .... 64 dBA
70 mph .... 69 dBA

MAINTENANCE
Oil/filter change ... 7500 mi/7500 mi
Tuneup .... 30,000 mi
Basic warranty ... 84 mo/100,000 mi

INSTRUMENTATION
150-mph speedometer, 7000-rpm tach, oil press., coolant temp, volts, fuel level

ACCELERATION
Time to speed	Seconds
0–30 mph	2.7
0–40 mph	3.7
0–50 mph	5.1
0–60 mph	6.7
0–70 mph	8.7
0–80 mph	10.9
0–90 mph	14.0
0–100 mph	17.3

Time to distance
0–100 ft 3.3
0–500 ft 8.3
0–1320 ft (¼ mi): 15.0 @ 94.0 mph

BRAKING
Minimum stopping distance
From 60 mph .... 155 ft
From 80 mph .... 267 ft
Control .... excellent
Pedal effort for 0.5g stop ...... 20 lb
Fade, effort after six 0.5g stops from
60 mph .... 25 lb
Brake feel .... very good
Overall brake rating ...... very good

FUEL ECONOMY
Normal driving .... 17.0 mpg
EPA city/highway .... 14/21 mpg
Cruise range .... 375 miles
Fuel capacity .... 23.0 gal.

HANDLING
Lateral accel (200-ft skidpad) .. 0.81g
Balance .... mild understeer
Speed thru 700-ft slalom .. 59.7 mph
Balance .... mild understeer
Lateral seat support ...... very good

Subjective ratings consist of excellent, very good, good, average, poor.

Test Notes . . .

■ The Allante's steering feels fuzzy between cuts through the slalom's cones. But during them, the outside tire loads up, and accurate feel suddenly returns.

■ Around the skidpad, the Allante's traction control limits how much power you serve up to the front wheels. Nevertheless, 0.81g isn't bad for a 3900-lb car wearing 225-mm-wide tires.

■ The Allante rarely exhibits evidence of torque steer. Nail it off the line and the steering can seem light, but at higher speeds, the Cadillac's tossable tail masks any predilection for heavy understeer.

drive 500SL. Response is predictable, though steering feel is fuzzy at the instant of turn-in. As the tire forces grow from increasing slip angles, steering sensation quickly builds. Then it's just a matter of the proper right-left-right-left steering-wheel rhythm through the cones. Yes, Allanté understeers and runs wide at extreme speeds. And, yes, the rear end will, eventually, come out. No, it doesn't do anything nasty like snap into oversteer if you lift off. Allanté is a well-mannered luxury car even when driven like a sports car.

Torque steer, rarely evident even with nearly 300 lb.-ft. of torque, appears when you are heavy into the throttle with one wheel significantly lower than the other. But it's a rare occurrence easily corrected with the variable-rate power-assisted steering.

At the back end of the car there's a new subframe-mounted independent suspension with upper and lower control arms, electronic dampers and a lateral link. Toe-in changes are said to be minimal during hard braking, so the previous tendency for the car to oversteer is eliminated. Also enhanced is high-speed stability. On mountain switchbacks during hard cornering, there isn't the sensation that the rear end is trying to beat the front into the next corner.

Overall, the handling is quite good, and it's balanced with an excellent ride. Isolation is in the very best American tradition as tar strips, rain gutters or expansion joints have little if any effect on ride.

This smooth ride benefits from a revised, electronically controlled damping system called "Road Sensing Suspension." RSS is able to change damping from soft to firm in one or more of the shock absorbers in just 10 to 15 milliseconds. The fast response time allows ride engineers to leave the dampers in the soft mode for a much longer period of time. When a wheel hits a typical pothole, for example, the damper settings can change from soft to firm as many as five times in one compression stroke of the damper, according to Cadillac.

The new suspension stands out for its impact and ride-harshness isolation, along with positive roll control during both high- and low-speed changes in direction. Some lessening of pitch and dive are also obtained as the dampers can go firm in the blink of an eye. The jingle-jangle feel and tire-whopping sounds of a traditional sports suspension with performance tires are almost eliminated. Magnificent cornering capabilities remain, far beyond the confidence of most drivers without an SCCA competition license.

Allanté's Pininfarina body is essentially unchanged from its 1987 introduction: austere angles and limited adornment that typify Italian design. The front air dam is now larger and closer to the ground for improved high-speed stability. The side windows are now one-piece and the outside mirror patches have been moved forward, with minor improvements in the driver's outward vision, along with better sealing and wind-noise reduction. The understated but elegant design has withstood the test of passing years.

Also continued, but less welcomed, is the troublesome so-called "Phase II" folding soft top. It retains its gas-spring cover opening and power-operated header latch, but removal of the top or its return to the well is still manual. The time and effort required may be minimal, but it's still irritating in a car of the class. The removable aluminum hardtop remains on the option list. And both tops have a glass rear window with electric defogging.

There's no switch that we can throw to peer into the future: How firmly is Allanté ensconced in GM's now-volatile cosmos? The long-standing commitment to use the car—sales be damned—as the technology leader for its premier division might evaporate by the time you read this. What won't evaporate is the resolve of Cadillac engineers who fight for technical respectability daily.

Allanté is a high-tech but certainly not a "why tech" car. Engineers can precisely define the reasons for, as well as explain the results of, every improvement. Our independent test numbers confirm the substantial improvements in the car. However, all the technology is hidden underneath the same body that bowed in 1987. Will the fashion-conscious buyers care about the changes? Will they even know of them? Yet the car is what matters in R&T's fundamental view of the vast automotive drama. And the new Allanté *is* capable of a technically superior performance.

■ A formal, rectilinear interior complements the Allanté's classy, unadorned sheet metal. Thankfully, the Allanté's dash is festooned with fewer buttons than when first introduced; conventional gauges, too, are an improvement over the original's analog/video-game hybrids.

CADILLAC CONNECTIONS

The world's longest assembly line: The Allanté doesn't just fly on the ground

BY LOWELL C. PADDOCK
PHOTO BY JOHN LAMM

SOME PEOPLE FIND inspiration in the shower, others while driving to work. For Cadillac engineer Warren Hirschfield, the ideal locus was an Italian hotel room in the spring of 1982.

At the time, Hirschfield and a small mission from Cadillac's engineering and design staffs were in Torino sizing up coachbuilders for a sports/luxury car project. The idea had been gestating at Cadillac since the Motoramas of the early Fifties, but low anticipated volume always squelched its realization. "We were stuck," recalls Hirschfield, "unless we could find somebody who could come up with affordable tooling costs and the expertise to build in such low volumes."

That somebody turned out to be Sergio Pininfarina, whose firm had been on cordial terms with Cadillac since the late Fifties when Sergio's father, Batista, produced a series of 200 Eldorado Brougham sedans. When Cadillac came calling in 1982, Pininfarina—whose work for Ferrari held particular respect among GM's design staff—was the logical candidate to develop and produce what would evolve into the Allanté.

A decade earlier, GM President Ed Cole had flirted with a radical new aircraft that could cost-effectively deliver both cars and freight. Although the program never flew, Hirschfield learned of its potential from the head of the Military Airlift Command, General Robert Husing, and the two ideas—Allanté and airplane—eventually came together in a 3300-mile assembly line connected by an Alitalia 747, otherwise known as the "Airbridge."

Despite the Italian connection, Allantés are actually conceived in Detroit as Eldorado-based underbodies, which are flown to Torino—along with several other U.S.-sourced parts—aboard the Airbridge. After shortening the underbody slightly, Pininfarina's role begins with the assembly and painting of bodyshells at its headquarters in Grugliasco. The bodies are then transferred to the San Giorgio trim plant outside Torino, where, alongside Ferrari Testarossas, Allantés are finished. When completed, the bodies are trucked to Torino's Caselle airport and loaded, 56 at a time, into the fuselage of the Airbridge.

After their 8-hour flight, bodies are delivered as needed to GM's Detroit-Hamtramck plant, where a surprisingly small corner is devoted to Allanté production. Using a series of wire-guided platforms, the powertrain and suspension components are married to the waiting bodies.

That the Allanté and the Airbridge now both function smoothly says much about the patience and persistence required to overcome distance, language and the early glitches that flawed the Allanté's rollout. "Both of us were new to the ultra-luxury segment," explains Andrea Pininfarina, the third generation to manage the company. "We had the experience of building Ferraris for the exotic-car segment, but the idea of a luxury car that could be used on a daily basis was something new to us."

Although shifting Allanté production to the U.S. was once rumored, Cadillac has stuck by both the car and the system used to create it, noting its role in paving the way for the new Seville and Eldorado, products that are changing the way consumers think about Cadillacs. "The Allanté was really the first vehicle that gave us the inspiration to break out of the mold," says Cadillac General Manager John Grettenberger. "It may not have been what it should have been when it was first launched, but Cadillac has never turned its back on it."

COMPARISON TEST

Cadillac Allanté, Jaguar XJS, Mercedes 300SL

Yeah, yeah. These convertibles don't make any financial sense. But, then, you don't buy an Armani suit to keep warm.

BY KEVIN SMITH

PHOTOGRAPHY BY RICK CASEMORE

You should probably be sitting down when you read what these cars cost. We all know that Cadillac, Jaguar, and Mercedes-Benz make interesting automobiles and that convertibles are always premium models. And there's no question that the Allanté, the XJS, and the 300SL have great appeal. But they all have serious shortcomings, too. We're not sure that should be allowed at these prices.

Look what the combined efforts of car-company marketers and government revenuers have done. Counting manufacturer's suggested retail price, freight, gas-guzzler penalty, and luxury tax, the Jaguar XJS convertible now costs nearly $72,000. And except for the folding top and numerous detail refinements, it is the same basic car the company introduced in 1975: elegant, yes, but also heavy, cramped, and thirsty. And it's still hobbled by the same slushy GM three-speed automatic transmission—which GM quit using generations ago.

The Mercedes-Benz 300SL is the newest design here and clearly the most modern. But it is also the most flagrantly priced car, at an astounding $90,335. That hit would perhaps be a little easier to take if a $12,000 Ford Escort LX-E didn't trounce the ostensibly sporty Benz by a full second in 0-to-60 and the standing quarter-mile.

Then there's the Cadillac Allanté, the bargain-basement ragtop of this group at $64,843. The new Northstar four-cam V-8 is a gem, but Allantés have been too expensive ever since the original was introduced at $56,000 (in 1986 dollars). Even if its competitors' price increases have somewhat outstripped its own, the Allanté forges on into the $60,000 bracket with its flexible structure and problematic top still intact.

Of course, these are flash cars, selected by emotion more than logic. But every judgment of an automobile must be made with one eye on the ask-

Jaguar XJS V12

Highs: Great to be seen in, rides like the *QE2*.

Lows: Too many details feeling too dated.

The Verdict: There will always be an England; this XJS, too?

ing price. With that firmly in mind, we set off with these exclusive convertibles: the new 1993 Allanté, with its 32-valve V-8 engine, friendlier seats, and revamped rear suspension (*C/D,* February 1992); the XJS, which sports some subtly revised sheetmetal and a reworked interior (*C/D,* July 1991); and the 3.0-liter six-cylinder version of Mercedes' SL two-seater, unchanged since its 1990 debut (*C/D,* September 1989). We gathered performance data at Willow Springs Raceway and ran up a zillion miles looping through the coastal mountains to Ojai and Santa Barbara before droning back to Los Angeles. Then we pondered the meaning of it all, and let the scores fall where they might.

And fall they did. Note that the Overall Rating numbers, in addition to clustering almost too close to call, barely broke out of the low eighties. The winning score was the lowest in recent comparison-test history. It's almost a shame to cite winners and losers under such circumstances, but here is our call.

Third Place
Jaguar XJS V12

Many of us have long had a place in our hearts for the velvety V-12 Jaguar sports tourer. The XJS enjoyed an extensive, if subtle, rework for 1992, incorporating new headlamps, taillamps, and grille, plus faintly bolder contouring of the rocker panels and hood bulge. (The coupe version also had its rear-quarter windows reshaped.) And a complete refurbish inside included new power-adjustable seats, a new wheel, and a new instrument panel with proper round gauges set in walnut.

Despite the upgrades, however, the XJS remains an aging design. Modern efficiencies—of space, weight, and energy—go unrecognized by this regal conveyance. The huge engine/transmission assembly squeezes passenger space, and the conversion to a soft-top transforms an already weighty two-plus-two coupe into a staggeringly heavy (4160-pound) two-seater—one that turned a gallon of unleaded premium into heat and smog every thirteen miles on our trip.

Does all this mean we didn't like driving the XJS convertible? Hardly. The big Jag rides beautifully, it has a sleek look and a stately demeanor, it makes respectable time over most roads, and it is in many ways the most flattering car of this group to be seen in. Though the steering feels more vague and slow than we like, the car can be guided through bends with fine balance—impressive, given the extremely soft suspension. And any time the tach is showing over 4000 rpm, the big V-12 can deliver good thrust. So certainly for posing, cruising, touring, and swooping through the mountains at a six-

Mercedes-Benz 300SL

Highs: Body rigidity, five-speed automatic, robo-top.

Lows: Too many bucks, too few horses.

The Verdict: Makes the $107,000 500SL look like a must-buy.

tenths pace, the XJS does the job.

But for 5.3 liters and a quoted 276 horsepower, the big twelve doesn't deliver much kick. Only the presence of the embarrassingly underpowered 300SL saved the XJS from the wrong kind of performance distinction in this test. And the V-12 gets little help from the Turbo-Hydra-matic three-speed automatic; this transmission works fine by decade-old standards, but today its shifts feel lazy and its selector lever balky.

The Jag scores pretty well as a convertible. It suffers only minor drafts in the open cockpit, and its easy-to-manage top mechanism—manual latches and power up-down, but a too-tight snap-on boot—is surpassed here only by the Mercedes'

The 500SL: pricier but more popular.

Normally, we would be the last people on earth to belittle $16,610. That sum will buy a fully loaded Nissan Sentra SE-R, with enough left over for a couple of years of insurance. It will pay for a year's tuition at Harvard. It is even, according to the government, sufficient annual income to keep a family of four off the welfare rolls. But when it comes to high-brow convertibles, we would spend that amount in a second, because $16,610 is what it takes to upgrade from a 300SL to a 500SL.

Generally, we love everything about the 300SL except its engine. The 3.0-liter six simply can't provide the 3980-pound convertible with the effortless thrust such a lavish machine deserves.

The 32-valve double-overhead-cam 5.0-liter V-8 used in the 500SL solves this problem. With 322 horsepower and 322 pound-feet of torque (94 hp and 121 pound-feet up on the six), the 120-pound-heavier V-8 model hits 60 mph in 6.3 seconds, covers the quarter-mile in 14.6 seconds at 99 mph, and pulls strongly to its 155-mph top-speed governor. Not only is the 500SL vastly more energetic than the 300SL, but after the first few feet from rest it also outruns the powerful Northstar-equipped Allanté.

More important than sheer speed, though, is the V-8's responsiveness. Stomp on the go-pedal and the 500 leaps forward. Ease into the throttle and it surges ahead in perfect synchronization, without any downshifting or hesitation.

Apart from the uprated engine, the 500SL is identical to its junior partner (with the exception of standard traction control, which is a $2822 option on the 300). But one's choice of SLs isn't about dollars and features. It's about making the SL's superiority complete and spending $106,945 instead of $90,335 to do it. Eight out of ten SL buyers think the 500 is the way to go. So do we.
—*Csaba Csere*

Vehicle type: front-engine, rear-wheel-drive, 2-passenger, 2-door convertible/coupe
Base price: $106,945
Engine type: DOHC 32-valve V-8, aluminum block and heads, Bosch engine-control system with port fuel injection

Displacement	303 cu in, 4973cc
Power (SAE net)	322 bhp @ 5500 rpm
Transmission	4-speed automatic
Wheelbase	99.0 in
Length	176.0 in
Curb weight	4100 lb
Zero to 60 mph	6.3 sec
Zero to 100 mph	15.1 sec
Zero to 120 mph	22.3 sec
Standing ¼-mile	14.6 sec @ 99 mph
Top speed	155 mph
Braking, 70–0 mph	175 ft
Roadholding, 300-ft-dia skidpad	0.82 g
Road horsepower @ 50 mph	14 hp
EPA fuel economy, city driving	14 mpg

ALLANTE/XJS/300SL

Cadillac Allanté

Highs: Great new powertrain, friendly handling.

Lows: Shaky structure, bland interior, fussy top.

The Verdict: Still pricey, but a decent value by comparison.

magical one-touch system. In body stiffness, though, the XJS takes the booby prize, flexing and shaking more than the others.

Frankly, for anyone who doesn't *need* to have a convertible, the XJS coupe makes a vastly better automobile, with much greater structural rigidity, more useful space, 200 fewer pounds, and a price tag $7000 friendlier.

Second Place
Mercedes-Benz 300SL

Hopping into today's Mercedes two-seater and feeling the solidity and quality, we could almost forgive Daimler-Benz for dragging the hallowed 300SL name down off its pedestal and thrusting it into the melee of the modern market. Then we stood on the throttle for a quick getaway and got all steamed up again. This thing should have been *at least* a 320SL, using the 3.2-liter six now fitted in the 300SE. That engine's fourteen-percent torque advantage would be a help because, as it is, the SL's 3980 pounds completely overwhelm the poor little 3.0-liter six.

Get past its acceleration rate, though, and the 300SL begins piling on rationales for its shocking sticker. The apparent stiffness of its body structure is remarkable, with only a rare hint of shudder in the cowl. The chassis is far and away the most advanced and best settled in this group, giving the SL a secure, confidence-inspiring chuckability over fast, undulating roads of any surface, as well as a fine freeway ride. Its comfortable seats—more like occupant-support systems, really—incorporate the whole safety-belt mechanism plus the usual power adjustments. Each unit probably costs and weighs as much as a Hyundai. Then there's the pop-up rollover bar, the clean body lines, the high quality of the finish—on and on.

And don't forget that top, the easiest and trickiest in creation. Hold a single button on the center console and watch with wonder as eleven solenoids, fifteen hydraulic servos, and seventeen proximity switches run their routine, throwing latches, raising panels, cranking the top as bidden, and buttoning everything down. Amazing. (Yeah, we know 1960s Lincolns

C/D Test Results

	acceleration, sec				street start, 5–60 mph	top gear, 30–50 mph	top gear, 50–70 mph	top speed, mph	braking, 70–0 mph, ft
	0–60 mph	0–100 mph	0–120 mph	1/4-mile					
CADILLAC ALLANTE (1993)	6.2	16.7	27.0	14.8 @ 95 mph	6.4	3.0	3.9	144	196
JAGUAR XJS V12	8.7	22.3	45.9	16.7 @ 87 mph	8.8	4.2	6.5	141	179
MERCEDES-BENZ 300SL	8.8	25.0	48.5	17.0 @ 85 mph	9.3	4.6	6.4	136	175

Vital Statistics

	price, base/ as tested	engine	SAE net power/torque	transmission/ gear ratios:1/ maximum test speed, mph/ axle ratio:1	curb weight, lb	weight distribution, % F/R
CADILLAC ALLANTE (1993)	$64,843/ $64,843 (estimated)	DOHC 32-valve V-8, 279 cu in (4565cc), aluminum block and heads, GM engine-control system with port fuel injection	290 bhp @ 5600 rpm/ 290 lb-ft @ 4400 rpm	4-speed auto/ 2.96, 1.63, 1.00, 0.68/ 45, 81, 132, 144/ 3.71	3720	62.9/37.1
JAGUAR XJS V12	$71,888/ $71,888	SOHC V-12, 326 cu in (5344cc), aluminum block and heads, Lucas engine-control system with port fuel injection	276 bhp @ 5550 rpm/ 306 lb-ft @ 2800 rpm	3-speed auto/ 2.50, 1.50, 1.00/ 68, 113, 141/ 2.88	4160	55.3/44.7
MERCEDES-BENZ 300SL	$90,335/ $90,335	DOHC 24-valve 6-in-line, 187 cu in (2960cc), iron block and aluminum head, Bosch KE-V Jetronic engine-control system with port fuel injection	228 bhp @ 6300 rpm/ 201 lb-ft @ 4600 rpm	5-speed auto/ 3.87, 2.25, 1.44, 1.00, 0.75/ 36, 63, 98, 136, 136/ 3.69	3980	51.3/48.7

did the same thing. But they didn't come with an automatic roll bar, did they?) The Mercedes also includes a removable hard top—an amenity that is $2000 extra with the Cadillac and not offered at all with the Jaguar.

And give credit where it's due: though the 300SL lagged behind the Allanté and XJS in all tests of acceleration and speed, it cleaned house in braking distance, skidpad grip, and fuel economy (according to both the EPA's test cycle and our trip log).

So there's no question the 300SL delivers some value. But in this comparison, and considering that lofty price, it wasn't enough to bag all the chips.

First Place
Cadillac Allanté

Cadillac's Pininfarina-made Allanté roadster has always seemed too posh to be a sports car but too brash to be a luxury cruiser. It targeted the two-seat Mercedes on price, performance, and accommodations but has never had the substance to go head-to-head with the high-quality, high-resale-value Benz.

A half-dozen years after its launch, however, the Allanté has inherited a huge price advantage. With the SL breaking through $90,000 (even in the "cheap" six-cylinder form tested here), the new Allanté's 1993 base price of $64,843—including the mandatory luxury and gas-guzzler taxes—suddenly looks almost reasonable.

But even bigger news for 1993, and the thing that makes the Allanté finally worth taking seriously, is the arrival of the much-anticipated Northstar V-8. This impressive new 4.6-liter powerplant brings Lexus/Infiniti-class technology, power, and smoothness to the domestic camp, even though the Allanté's Italian body assembly gets it classed an import. This brilliant all-aluminum multivalver delivers 290 horsepower through a slick and responsive new four-speed automatic.

In every contest of speed and response, Cadillac's Allanté has it all over the Mercedes 300SL and the Jaguar XJS. Punching the Northstar's throttle to leap into a hole in traffic on Santa Barbara's crowded State Street, or to launch out of a slow bend on serpentine Highway 33, always brought an effective boot in the back. And a nice exhaust growl, too.

roadholding, 300-ft skidpad, g	interior sound level, dBA				fuel economy, mpg		
	idle	full throttle	70-mph cruising	70-mph coasting	EPA city	EPA highway	C/D 400-mile trip
0.77	48	78	74	73	14	21	16
0.73	47	75	73	73	12	16	13
0.80	46	77	74	74	16	23	19

dimensions, in				fuel tank, gal	interior vol, cu ft		suspension		brakes, F/R	tires
wheel-base	length	width	height		front	trunk	front	rear		
99.4	178.7	73.4	51.5	23.0	55	16	ind, strut located by 1 lateral link and 1 trailing link, coil springs, electronically controlled shock absorbers, anti-roll bar	ind, unequal-length control arms, 1 toe-control link, coil springs, electronically controlled shock absorbers, anti-roll bar	vented disc/ disc; anti-lock control	Goodyear Eagle GA, P225/60ZR-16
102.0	187.6	70.6	49.4	23.5	50	9	ind, unequal-length control arms, coil springs, anti-roll bar	ind; fixed-length half-shaft, 1 control arm, 1 trailing link, and 2 coil-shock units per side	vented disc/ disc; anti-lock control	Goodyear Eagle NCT 60, 235/60VR-15
99.0	176.0	71.3	51.3	21.1	50	8	ind, strut located by a control arm, coil springs, anti-roll bar	ind, 2 lateral links and 3 diagonal links per side, coil springs, anti-roll bar	vented disc/ disc; anti-lock control	Pirelli P600, 225/55ZR-16

ALLANTE/XJS/300SL

Though 290 horsepower sounds like too much to send through front wheels, the Allanté's new chassis manages the challenge just fine. There are reworked struts in front and unequal-length arms replacing the former struts in back, plus variable damping that reads and reacts to wheel travel in quasi-active fashion. The car does not have the sense of near-perfect balance and fine damping of the 300SL, and it certainly doesn't have the structure, but the Allanté feels easy, natural, and willing when the corners are coming up thick and fast. Stopping distances with our test car were on the long side, due perhaps to tire choice or the car's extreme forward weight bias. But otherwise, the sportiest Cad performed with distinction.

Even if it scored well enough to win here, though, the Allanté still gives us some things to wonder about. Such as, do its customers really like an instrument panel that's all straight lines, right angles, and fields of identical buttons? Is its interior trim rich enough for this price range? Couldn't the structure be tightened up? Don't the proprietary GM ignition switch and transmission selector feel sloppy in a $60,000 automobile? And why did the manually operated, power-latched top fit as if it were a half-size too small, forcing us into energetic tricks to get both ends attached?

Ah, well, maybe we're being too hard on these things. Cars like these may come up short on normal automotive value, but they aren't really normal automobiles. These are boutique cars, purchased largely as fashion accessories. You don't buy an Armani suit to keep warm.

The reality is, an open car will always be more complex and costly to create, and it can hardly avoid being heavier, less sturdy, less efficient, and less comfortable as a result of its convertibility. That hinders the car's effectiveness, but it isn't really an argument for eliminating the open-air option. There is something undeniably opulent about an expensive car with its top folded away and its occupants on display.

For the right people (wealthy, and willing to look it) and the right missions (tennis, lunch, shopping, or a cruise down the Coast Highway to Sunday breakfast), the Allanté, the XJS, or the 300SL could work just dandy. But you'll just have to forgive the manufacturers if the bold prices of this trio seem to promise a broader repertoire than that. •

Editors' Ratings	engine	transmission	brakes	handling	ride	ergonomics	comfort	panache	value	styling	fun to drive	OVERALL RATING*
CADILLAC ALLANTE (1993)	9	9	8	8	8	8	8	7	8	8	8	84
JAGUAR XJS V12	7	6	8	7	9	7	8	9	6	9	7	79
MERCEDES-BENZ 300SL	7	8	9	9	8	9	8	9	6	9	8	82

HOW IT WORKS: Editors rate vehicles from 1 to 10 (10 being best) in each category, then scores are collected and averaged, resulting in the numbers shown above.

*The Overall Rating is not the total of those numbers. Rather, it is an independent judgment (on a 1-to-100 scale) that includes other factors—even personal preferences—not easily categorized.

DRIVING THE NEW CARS

'NORTHSTAR' ALLANTE

The injection of a brawny new V8 has made Cadillac's two-seater roadster the most powerful front-driver in series production world-wide. With new 'road sensing suspension', it might even challenge the imported sports cars, on merit. But will it? Not likely...

REPORT BY CHRISTOPHER JENSEN/JUKKA SIHVONEN

ALTHOUGH the Cadillac Allante's body is virtually unchanged this year, the two-seater is undergoing a spiritual transformation that should appeal to those who favour rapid-transit luxury roadsters. The 1993 Allante (which went on sale in the United States in April and is due to reach Europe in September), is powered by Cadillac's new 217 kW, 32-valve V8 known by the codename "Northstar". That makes it the most powerful production front-wheel driver in the world.

In addition, the interior has been improved and more important, its adaptive suspension has gone from rube to sophisticate – an evolutionary advance of which even Charles Darwin would approve. Cadillac officials are so pleased with the Northstar Allante's performance that they swear (cross their little corporate hearts and hope to die) that a completely stock version, aside from safety equipment such as a roll-over bar, will be used as the pace car at the next Indianapolis 500 race.

Those changes may not seem terribly important to those not interested in a two-seater expected to sell for about R190 000 in the United States. But

Cadillac engineers say the new V8 and "Road Sensing Suspension" will quickly trickle down into less-expensive models such as the Seville sedan, which is also entering the European market.

Introduced as a 1987 model, the Allante was a mating of a Pininfarina body and interior and American mechanicals. But the car has not had much success. Cadillac has had a tough time convincing the wealthy to try their roadster, when their arch-competitor Mercedes has been offering an established quantity in its "SL" sports car range.

Cadillac hopes that the Northstar V8 and the new suspension will resuscitate the Allante in the United States and Europe, where Cadillac admits that only a "handful" of previous versions have been sold. The defibrillator for that revival is the engine, which has been developed by Cadillac and will remain exclusive to the marque for at least a year.

Externally, the only change to the Pininfarina body design is a 75 mm deep air dam. Inside there are a few more changes, including new seats that are easy to adjust and extremely comfortable to use.

When buttons die...

The Allante's "driver information centre" is angled nicely towards the driver and provides all the proper information. But it seems awfully busy, as if this is where buttons go to when they die.

The questionable styling of the Allante has benefited from the new, deeper air dam since its 1987 launch (above). And now, with 217 kW of new wave Detroit muscle (below) it reaches 100 km/h from rest in a claimed seven seconds.

The controls are fairly easy to use but the button-mania seems daunting, just at first. On the other hand, the switchgear's feel is better than that found on many European and American luxury cars, although not quite up to the standards set by cars like the Toyota Lexus, the Nissan Infiniti and the Honda Acura.

One feature that will never be found on a Mercedes 500 SL is an information panel that greets you when the Allante is started, choosing its salutation according to the time. "Good morning," it says – and you soon expect it to start offering advice... "Good morning. You are not going to wear that tie with that shirt, are you?"

There is a small storage space behind the seats and the boot will hold about 368 dm³ of your worldly belongings, which is not bad for a roadster. The Pininfarina-designed, manually-operated soft top has also been reworked. It

The driver information centre above the console is crammed full of buttons and takes getting used to. Other technology includes computerised "adaptive suspension".

was a disaster when introduced and Cadillac officials confide they put too much trust in Pininfarina's engineering capabilities.

It was hard to use and overly complicated. Missing one step would allow it to jam in the down position, raising the possibility that the Allante would become an automotive counterpart to a Waterford goblet.

The revised top is now acceptable, although the lining should be improved to reduce wind noise. An aluminium hardtop is also available.

The important feature is the transversely mounted motor, a 4,6-litre, d-o-h-c, 32-valve, 90° aluminium V8. Running on 10,3:1 compression, it's rated at 217 kW at 5 500 r/min with 393 N.m of torque at 4 400.

The Northstar is Cadillac's first all-new engine in more than a decade and it replaces an old push-rod 4,5-litre that was rated at 150 kW at 4 400 r/min and 366 N.m at 3 200 revs. The engine is coupled to a new, electronically controlled four-speed automatic transmission developed specifically for Northstar. No manual gearbox is available.

The result is a package as slick and flexible as an American presidential candidate at full campaign speed, presenting a solid challenge to the "guts-and-cream" 4,0-litre V8 used in Toyota's Lexus LS 400. Even Cadillac officials, however, admit it is marginally behind the Mercedes 500 SL's V8 in both smoothness and response.

233 km/h top speed

The Allante pulls away from a dead-steady idle with an even, steady progress barely broken by upshifts that are nicely slurred, as a torque-management computer retards the timing at the proper moment. Furthermore, while offering a pleasingly large portion of low-speed torque, the Northstar doesn't go breathless at higher engine speeds.

My drive in a pilot car did not allow for instrumented testing but Cadillac says the latest Allante (kerb weight 1 711 kg) will accelerate from 0 – 60 mph (96,6 km/h) in seven seconds and achieve a top speed of "over 233 km/h".

Despite the 217 kW output, there is virtually no torque steer: an occasional tug here and there but nothing naughty. There is also a traction control feature, to help put down the power on a slick surface.

Happily, considering the new-found muscle, the revamped suspension works well. The rear suspension gets a new short/long arm, multi-link arrangement to replace the old struts, which improves stability and minimises body lift under hard braking.

More important is the adaptive suspension. The old Allante had a speed-regulated, three-stage system in which all four corners dutifully went from "comfort" into "normal" at 65 km/h, switching into "firm" at 100 km/h. This was not a system that was likely to keep Mercedes engineers awake at night, unless they had giggling fits.

In contrast, the new Road Sensing Suspension is a vast improvement. Its sensors include one at each corner, four accelerometers and the engine-management computer to provide information such as wheel position, lift and dive. What the system lacks, however, is input directly from the steering.

The information is assessed by the RSS computer which reacts by ordering a solenoid valve on the front strut (or rear shock) to move into either a firm or soft mode. Cadillac engineers say the response time at the shock is 10 to 15 milliseconds and the result is a kind of "real-time" damping. The system can open and close the valve several times as the wheel travels through one undulation, for example.

All this frantic activity is invisible to the driver. The car simply seems eerily composed. On a smooth surface, the ride is gentle without feeling either light or floaty.

Passive but good...

Surprise changes in the road surface and manoeuvres such as dips, off-camber turns, lifts, dive and body lean are dealt with easily. The Allante just feels a little tighter, momentarily, before relaxing again. It performs at its best when absorbing really big inputs such as those from major swells in the tarmac. Sharper inputs such as potholes are still felt by the car's occupants. This is, after all, still a passive system...

A reasonable argument can be made that the new Caddy system is more pleasant than the so-called "fully active" suspension found on Nissan's Infiniti Q45. The Q45 deals with body motions in a no-nonsense fashion, but the downside is a constant firmness that often verges on being harsh.

Cadillac has improved the speed-sensitive steering, too. While the old system was not bad, it did have some vagueness on centre point and it was limited in its range of adjustment.

The new system tidies up the on-centre slop and the speed-sensitive adjustments do a far better job of spanning the parking lot-go-directly-to-jail range, continuing to tighten things down right from 30 to 205 km/h. From there on, the effort is steady.

The result is that the Northstar Allante feels terribly friendly and reassuring. The body rigidity is good (though not as good as that of Merc's 500 SL) and it is very easy to go quite quickly on anything but a wickedly serpentine road.

Where twist comes to turn is where the Allante shows its front-wheel drive nature. Although neither ponderous nor slow to turn in, despite 63 per cent of the weight being up front (up from 61 per cent with the older car), there is a slight pause before the Allante responds and some drivers are still likely to yearn for such rear-wheel drive placement options as lift-throttle oversteer.

Nevertheless Cadillac thinks consumers will find the front-wheel drive benefits of improved traction (in some cases) and packaging are well worth it. The 1993 Allante is an entertaining and comfortable car but at this price level, it might take a buyer with self-confidence not to make the more obvious, hotsy-totsy choices from among the imported marques. ●

COMPARISON
Prime Rib AND SUSHI

AMERICA VS JAPAN — THE BATTLE OF THE MULTI-VALVE V8 LUXURY LINERS

STORY AND PHOTOGRAPHY BY DAVID FETHERSTON

You've read about trickle-down economic theories in the news for the past 10 years — this is supposed to mean that as the rich make money, the rest of us get an improvement in our economic growth due to their business expansion. As the US economy has so perfectly proved, trickle-down economics doesn't work.

This same theory has been used by carmakers for decades and here it does work, but not in the same sense. In the car industry it "radiates" rather than "trickles down".

Detroit has been backed into a corner by foreign competition. The image of the all-encompassing Japanese car giants gobbling up the US car market in the '80s had a negative effect on everything — from corporate investment to worker morale. For many years Detroit's Big Three built cars just as they imagined buyers wanted them, but they discovered this was not the way to more sales. Meanwhile, Japanese manufacturers were surpassing Detroit by leaps and bounds.

Detroit (aka Motown — "motor-town") has been the home of the car since the beginning of this century but, as the '80s approached, it was obvious that the Japanese were building cheaper, yet more efficient and better designed, cars. Many of them were also more powerful.

But as sales in the late '80s and early '90s have proven, the US car industry is working its way back to a position of power. It is surprising to learn just how much Detroit's cars have improved due to this pressure. Indeed, on both sides of the Pacific the intense competition has seen ever-more sophisticated technology radiate throughout product ranges, in everything from small, hi-tech four cylinder screamers, to a new generation of multi-cam, multi-valve, V8 engines gracing the engine bays of the upper luxury league — cars such as Cadillac, Lincoln and Japan's Lexus (Toyota) and Infiniti (Nissan).

Multi-valve technology has been with us for years, first of all on the race track. It came into vogue on street cars after turbo engines had their day. Not only did they offer generous power, they ran cooler and produced a cleaner exhaust. Audi and Toyota were among the first to introduce a modern V8 engine using this same thinking. Toyota's effort resulted in its wonderful Lexus line, featuring a 190 kW, 4.0-litre, multi-cam, multi-valve V8.

It sold like hot cakes. In the US, Lexus LS400 grabbed a large slice of the luxury market from Cadillac, Mercedes, BMW, Audi, Lincoln and Jaguar. Its impact had a dramatic and polarising effect. It revved up a few programs which were shuffling along in the corporate development stages. One of those was the long awaited Northstar V8 from Cadillac, the first of the new Detroit V8 engines to hit the street. Ford will slot its own hi-tech V8 in the Lincoln Mark VIII this month.

Cadillac's Northstar V8 made its debut in the '93 Cadillac Allante sports. So we've decided to see just how well Detroit's new tech compares to Tokyo's hi-tech. We chose the sporty new Lexus SC400 Coupe to pair off with the Allante, based on engineering rather than simple seating capacity.

The SC400 has already proved its worth as one of the finest luxury sports coupes of the '90s. It comes with more bells and whistles than the Vatican has priests, as

On the twisty stuff the SC400 does a brilliant job of going point-to-point

much luxury and comfort as a five-star hotel and enough power to take you into "ticket territory" in less than eight seconds.

The Cadillac Allante has been part of the Cadillac line-up since '87. Designed by Pininfarina on commission from Cadillac to add a little European flavour to its range, until the '93 model year the Allante was powered by the 4.9-litre aluminum V8 used in the regular Cadillac sedans and coupes. With only 149 kW on tap the Allante was not the sports car it should have been. Now it comes with the Northstar 216 kW multi-valve V8.

In many ways the SC400 and the Allante match each other point for point. They are of similar size and weight with power-to-weight ratios that run 14.7 kg/kW for the Lexus and 12.8 for the Allante. One significant difference in the Allante is its front-wheel drive; the Lexus is rear drive.

When you sit down and do the numbers they produce surprisingly similar acceleration. The SC400 and the Allante both run 0-100 km/h in 6.7 seconds and cover the standing 400 metre in 15.2s (149.6 km/h) and 15.0s (150.7 km/h), respectively. While top speed was not tested, both makers claim 233 km/h which I guess makes them kissing cousins in the performance game.

Designed in California at Toyota's Calty Design Research facility, the SC400 personifies the modern age and grace of the "automotive egg-look." Its wonderful rounded lines look like they have just popped from a show car in the year 2000. However, this car is as real as they come and riveting to drive. The styling is uniquely Lexus and finely done without any compromises that hinder or break its organic lines. The detailing of the panel fit and finish is what the Europeans dream about and only Mercedes comes close to.

Every break line in the body is a perfect fit. The side panel has one crease which gives it some squareness to break the roundness. Even the bonnet has two fine contour lines which run down into the nose inboard of the head lamps. A plastic nose and tail cone encapsulate both ends.

When I first drove the SC400 a year ago it immediately reminded me of a 928 S4 Porsche. Its "fat" feel on the road, its sumptuous leather interior and the heavily racked windshield floating over the wide dash, blend into a roomy and luxurious cockpit. The space is immediately welcoming. You know you are in a vehicle that is designed and assembled with care and craftsmanship. In some ways it's a bit like coming home to a comfortable living-room.

The interior's muted tones give it an air of class and quality. The seemingly over-stuffed armchair bucket seats allow the body to settle down comfortably. They surround and support the rump, legs and back for long distance "flight". The dash is two-toned with a bird's-

eye maple trim, its shapes mimicking the rounded exterior and flowing into the pod-like air-conditioning/sound system panel which sits above the full width console and armrest.

The steering wheel and hub features an SRS air bag and cruise control as standard, as well as a combination two-memory seat and steering wheel position function. The steering wheel rises away from the driver immediately the ignition is turned off, as it does in the big brother LS400, to help entry and exit, and the steering column has a four-way power adjuster as well. This places the wheel just perfectly at your fingertips.

The rear seats are unlike the 928 — they can actually seat two adults in comfort.

The dash uses one of Lexus's interesting electro-luminescent displays which light up when the engine is in operation. It is not digital but rather a full analogue design with layered panels of colour and lighted needles.

The true test is in its driving and, as I found, the SC400 can turn up the heat and crack along so far over the legal limit that its performance envelope is hardly in question. Surprisingly, the coupe is 2 km/h slower than the sedan, a point you wouldn't want to argue with the local magistrate.

On the twisty stuff the SC400 does a brilliant job of going point-to-point. Surprisingly for a big car it doesn't mind being thrown around too hard. The suspension uses an unequal length control arm arrangement up front with a more complex trailing link, lateral link and control arm design on the rear.

The layout is well tuned to the weight of the coupe and provides a balance that takes a lot of pounding before anything less than full control is even hinted at. It all inspires confidence in the car's dynamic control, something underpinned by the combination of traction control (optional) and ABS brakes. Running with the pack is easy and outpacing it is just a breath away. Just like the 928, the SC400 needs room to breathe and once it gets a lung-full it takes on a new character of speed and grace.

Power is produced so smoothly and without fuss that the car's actual road speed can be deceptively quick. When the LS400 was introduced in California, they had a sedan up on stands.

Running the engine at maximum rpm, an engineer placed a wine glass on top of the induction manifold. The wine had hardly a ripple on the surface. Smooth operator for sure!

The rousing power and torque figures for this 4.0-litre V8 are 190 kW and 360 Nm. Its design and fabrication is exquisite. It features an aluminium/steel liner block with four-valve, dual-cam heads running 10.0:1 compression. Fine points, such as the nearly silent hydraulically-driven radiator fan, are missed by most but when it gets into a full

Cadillac Allante (LEFT) looks staunchly American but in fact was penned by Pininfarina. Not too many curves to be seen in the Allante's interior, in keeping with expectations of America's renowned luxury marque. The power behind the scenes is the Northstar 32-valve V8, which with 216 kW on tap replaces the 4.9-litre aluminium V8 in the previous Allante, which offered only 149 kW.

Lexus SC400 (RIGHT), designed at Toyota's Calty centre in California, sports elegantly modern lines. Interior (BELOW) exudes quality — sumptuous leather seats and maple wood trim completes the impression of understated modern luxury. Four-litre V8 is one of the smoothest engines in the business.

wail the quiet power delivery is nicely trumpeted by the sporty exhaust tune.

At its $US38,600 base price, the SC400 is actually cheaper than the LS400 sedan. But add a couple of extras including traction control and you have $US42,200.

At that price it's a bargain especially in light of the fact that the Porsche 928 GTS demands $US94,000, the Allante is $US65,000, the Mercedes 560 SEC is $US90,000 and the BMW 850i coupe is $US80,000.

This is a coupe you can hardly fail to love and enjoy. It is a delight to drive; it rides like a good car should and handles the twisties with enough adhesive power to easily scare your neighbours without breaking the law. If you do want to run with all the big boys, the 233 km/h top speed is fast enough . . .

Just as the Japanese came to California to have their SC400 designed, so Cadillac went to Italy in '82 to have the Allante penned. Cadillac wanted a European flair for the new sports so it was handed to Sergio Pininfarina to pen the new roadster. It also continues the long term relationship the company has had with Cadillac. In early '59 and '60 Pininfarina built several hundred four-door Eldorado Broughams under a similar arrangement.

The Allante is a fine and sophisticated design which exhibits a gracious elegance defining its own persona, while keeping the Cadillac corporate image glowing.

An Allante starts off in Detroit as an Eldorado floorpan. The pan is shipped to Italy where Pininfarina builds the Allante body, convertible top, fits the interior and electrical harness and paints it. The bodies are then shipped to Detroit from Turin, Italy, aboard a special Alitalia 747 air freighter, 56 at a time. Once they hit Detroit they go to Cadillac's assembly plant for powertrain, suspension and the rest of the mechanicals. The build quality is excellent. The body lines are well defined and true while the paint possesses a fine quality that one would expect on a $60,000 car.

It is a big car for a two-seater. Its wheelbase of 2286 mm is matched to a track width of 1524 mm to carry the 1676 kg body. The new convertible top mechanism is not the simplest but it does work fine and gives an airtight fit. When the Allante was introduced it came with a hardtop. These days it's a $US1600 option. I personally like the hardtop; it gives the roadster a more defined line, more in tune with the Cadillac image.

The interior enunciates the luxury image of Cadillac. Like the SC400 the leather seating is armchair style and inviting. Thankfully the original "games arcade" dash has been replaced by a simple analogue gauge cluster backed up by a series of warning lights, information windows and chimes. In the traditional sense, the dash does not carry off the sportscar image so well. But its 128 kilobyte memory in the powertrain computer produces a wealth of information on everything from fuel consumption to oil condition and maintenance schedule.

The computer also has self-diagnosing cycling which automatically checks everything from the fuel injection software to the tail lights, to fuel and coolant level. Most of this information is displayed in a window in the centre console. The steering wheel is also fitted with an SRS air bag which in these days of advanced tech now seems so ordinary; that is, until you have your first inflation. It's then that you thank your lucky stars . . .

From the driver's seat the Allante is not so much a sports roadster as the Lexus SC400 is a sports coupe. The Allante is rather a very good compromise between the two — a sporty handling luxury roadster. The

	Cadillac Allante 4.5-litre, 4-sp auto	Lexus SC400 4.0-litre, 4-sp auto
ENGINE		
Location	front, longitudinally mounted	front, longitudinally mounted
Cylinders	V8	V8
Bore x stroke	93.0 x 84.0 mm	87.5 x 82.5 mm
Induction	electronic multi-point port fuel injection	electronic multi-point port fuel injection
Compression ratio	10.3:1	10.0:1
Valve gear	double ohc/bank, four valves/cyl	double ohc/bank, four valves/cyl
Power	216 kW @ 5600 rpm	190 kW @ 5600 rpm
Torque	393 Nm @ 4400 rpm	360 Nm @ 4400 rpm
SUSPENSION		
Front	independent by MacPherson struts, lower A-arms and anti-roll bar	independent by MacPherson struts, unequal length control arms, and anti-roll bar
Rear	upper and lower A-arms, coil springs, toe links and sway bar	independent by coil springs, semi-trailing arm, two lateral links and anti-roll bar
Wheels	alloy, 7.0 x 16	alloy, 7.0 x 16
Tyres	P225/60 ZR 16	P225/55 VR 16
BRAKES		
Front	275 mm discs	297 mm discs
Rear	280 mm discs	280 mm discs
Anti-lock	yes	yes
STEERING		
Type	power assisted rack and pinion	power assisted rack and pinion
DIMENSIONS (mm)		
Wheelbase	2524	2690
Front track	1534	1521
Rear track	1534	1524
Overall length	4538	4851
Overall height	1308	1336
Kerb weight (kg)	1694	1673
Weight to power (kg/kW)	7.8	8.9
Fuel tank (l)	87.0	78.0
ACCELERATION (seconds - claimed)		
0-100 km/h	6.7	6.7
Standing 400m	15.0 seconds	15.2 seconds
Terminal speed (400m)	151 km/h	149 km/h
PRICE		
LIST PRICE	$US61,675	$US38,600

Cadillac's pleasures come wound up as a package of luxo-ride and sporty handling, neither of which is perfect but both delightful when used in their elements.

The suspension is all fresh on the '93, featuring RSS (Road Sensing Suspension), an electronic whiz able to change the shock absorbers' damping rate in less then 15 milliseconds, adjusting the suspension rate over surface variations instantaneously.

The suspension units are simple enough. The front uses a tower filled with a MacPherson strut, lower A-arm and a sway bar. The rear uses a subframe mounted independent suspension with upper and lower control arms, a lateral link and electronic dampers.

In many ways this electronic suspension management system is a form of stabiliser. It evens out the vehicle's pitch and rebound into a smooth, continuous motion much like a semi-active suspension. In demanding conditions the new suspension is right at home. With the 60/40 front and rear weight balance one would think that the tail would want to lead out of some corners but Cadillac has been able to eliminate that feel with this neat engineering.

The handling reveals mild understeer and unless you have a hankering to explore the edge of the performance envelope you will find the Allante a very satisfying drive.

The 4.5-litre Northstar V8 is one of the great engine designs. It uses the latest electronic tricks, fuel injection and healthy breathing, to produce 216 kW. Its vibration level is akin to the Lexus as is its lack of mechanical whirs and clicks. Its basic mechanical design features a two-piece block using an upper cylinder section and a lower cradle to carry the bottom half of the main bearings.

The heads use a chain driven, dohc, four valve hemi-head design. The heads and the chain drive are encapsulated with unique laminated steel covers which insulate the noise produced by the chain-to-sprocket and the valve gear. Ignition is direct, fired by a magnetic trigger off the crank, featuring four coils and no moving parts.

The induction system uses thermoplastic manifold pieces to save weight and improve flow. Fuel distribution is controlled by an electronic-sequential-port-fuel injection system governed by the 128 kilobyte engine computer. Maximum shift point is 6600 rpm which allows for some fairly snappy driving. Off the line the Allante does not produce the usual front drive/nose up weight transfer, and screaming tyres; the traction control and the engine computer take command and the only sensation you get is rapid forward motion. It does however have a little tyre chirp built into the program, to add a note to the sensation of power.

Take the power of the Northstar V8, the superb grip of the P225/60ZR-16 Goodyear Eagle GA tyres, the slick speed sensitive steering, four-channel ABS brakes, the standard traction control, the Allante's refreshingly low ride harshness and you have a handling package akin to Velcro.

The Allante is a swift and powerful car, with rocketing performance which is delightful, especially in light of how it gets up and dances so gracefully at speed. It is a more gutsy car to drive than the SC400, but that is not to demerit the SC400, which is a superb car.

But the Allante shows Detroit is once again a force to be reckoned with. Cadillac may have taken a few years more to make the Northstar just right but the results prove that it's not just the Japanese or Europeans who can build a decent engine and put it into a fine car; the boys in Detroit are ready to play major league in the '90s with new team members who can hit home runs. **M**

CADILLAC ALLANTÉ

For 1993, Cadillac's flagship convertible continues as the company's rolling showcase for new technology. This luxury roadster out of Detroit—by way of Pininfarina in Italy—enters the new model year equipped with nothing less than a new engine, new transmission, new suspension setup, a pot full of other high-tech goods and low-tech necessities like a dual cupholder in the center armrest.

At the head of this heady list of new features is the 295-bhp 4.6-liter Northstar V-8 engine. This powerplant is available in other Cadillacs in 1993, but Allanté had it first when it was introduced in January of 1992 as a 1993 model. This new source of power is combined with an all-new electronically controlled 4-speed automatic.

Allanté's status as one of the best-balanced front-drive cars in the world is enhanced by a number of new elements, including the Short/Long Arm (SLA) multi-link rear-suspension layout, which joins the Allanté's original MacPherson-strut front suspension.

Road Sensing Suspension (RSS)—yes, it's new—replaces last year's Speed Dependent Damping system. RSS manipulates the Allanté's ride using wheel-position and vehicle-body-motion sensors to determine how soft or how firm the damping should be.

A quicker-responding traction control system—capable of cutting off fuel to five of the engine's eight cylinders when the grip starts to slip—is also a part of the 1993 Allanté package. So, too, are larger rotors and calipers on the 2-seater's 4-wheel disc brakes.

What with all the neat technical features added to the new Allanté, it would be easy to neglect the tasteful enhancements to the car's exterior and interior. These include newly styled wheels—chromed if you wish—wrapped with Z-rated Goodyear Eagle GA tires. Three new exterior colors—Pearl Red, Pearl Flax, Polo Green—are added.

On the inside, driver and passenger are contained in leather-trimmed, orthopedically designed bucket seats with 8-way-position and 4-way-lumbar electric adjustments. For the audiophile, a new 8-speaker audio system is also standard.

As it has done in years past, Cadillac offers an optional removable (bring a friend) aluminum hardtop for the Allanté, featuring a rear-window defogger. The soft top, which is still a manual affair, also has a glass rear window with defogger.

Indeed, it is refreshing to talk about General Motors cars in terms of technology again. And as far as the Allanté goes, it'll talk your ear off.

SPECIFICATIONS

Base price, base model............ $61,675	Fuel capacity................. 23.0 gal.	Brakes, f/r.............. disc/disc, ABS
Country of origin/assembly....... U.S.A., Italy	Fuel economy (EPA), city/highway... 14/21 mpg	Tires.................... P225/60ZR-16
Body/seats..................... conv/2	Base engine......... 295-bhp dohc 32V V-8	Steering type............. rack & pinion (p)
Layout........................... F/F	Bore x stroke............. 93.0 x 84.0 mm	Turning circle................ 40.2 ft
Wheelbase.................... 99.4 in.	Displacement................. 4565 cc	Warranty, years/miles:
Track, f/r................. 60.4/60.4 in.	Compression ratio............... 10.3:1	Bumper-to-bumper........... 4/50,000
Length...................... 178.7 in.	Horsepower, SAE net... 295 bhp @ 5600 rpm	Powertrain............... 7/100,000
Width....................... 73.4 in.	Torque........... 290 lb-ft @ 4400 rpm	Rust-through............. 7/100,000
Height...................... 51.5 in.	Optional engine(s)................. none	Passive restraint, driver's side......... airbag
Luggage capacity............... 16.3 cu ft	Transmission..................... 4A	Front passenger's side........... none
Curb weight................... 3765 lb	Suspension, f/r................. ind/ind	